Ivo Senjanović
Marko Tomić
Neven Hadžić

Moving load and beam response

AF190943

Ivo Senjanović
Marko Tomić
Neven Hadžić

Moving load and beam response

**A systematic investigation and analytical
solution of Timoshenko beam vibrations due to
moving gravity and inertia force**

LAP LAMBERT Academic Publishing

Imprint

Any brand names and product names mentioned in this book are subject to trademark, brand or patent protection and are trademarks or registered trademarks of their respective holders. The use of brand names, product names, common names, trade names, product descriptions etc. even without a particular marking in this work is in no way to be construed to mean that such names may be regarded as unrestricted in respect of trademark and brand protection legislation and could thus be used by anyone.

Cover image: www.ingimage.com

Publisher:
LAP LAMBERT Academic Publishing
is a trademark of
Dodo Books Indian Ocean Ltd. and OmniScriptum S.R.L publishing group

120 High Road, East Finchley, London, N2 9ED, United Kingdom
Str. Armeneasca 28/1, office 1, Chisinau MD-2012, Republic of Moldova, Europe
Managing Directors: Ieva Konstantinova, Victoria Ursu
info@omniscriptum.com

Printed at: see last page
ISBN: 978-3-659-62951-8

Table of Contents

1 Introduction 3

2 State of the art 5

3 Motivation and research outline 9

4 Application of the Timoshenko beam theory 11

   4.1 Differential equation of vibrations 11

   4.2 Modal differential equations formulated by the Galerkin method 13

   4.3 Modal differential equations formulated by energy balance 15

5 Application of the modified Timoshenko beam theory 18

   5.1 Differential equation of vibrations 18

   5.2 Modal differential equations formulated by the Galerkin method 19

   5.3 Modal differential equations formulated by energy balance 20

6 Comparison of different formulations of differential equations 23

   6.1 Application of the Timoshenko beam theory 23

   6.2 Application of the modified Timoshenko beam theory 24

7 Forced vibrations excited by moving gravity force 26

8 Forced vibrations excited by moving gravity and inertia force 30

   8.1 Modal inertia forces 30

   8.2 Modal differential equations 31

   8.3 Solution of linear vibrations 33

   8.4 Solution of subresonant parametric vibrations 35

   8.5 Solution of resonant parametric vibrations 38

      8.5.1 Diagonal parametric time functions 38

      8.5.2 Off-diagonal parametric time functions 40

9 Beam vibrations after the passing of moving mass 43

10 Numerical examples 44

11 Conclusion 54

1 Introduction

Beam-like engineering structures are ordinarily analyzed in the early design stage, when all structural elements are not yet defined, as a beam. Hence, instead of a 3D FEM model, a 1D FEM model is used with cross-section properties determined as equivalent quantities of the 2D sectional structure. In the case of structures with a large aspect ratio of height and length, the Timoshenko beam theory is used, which takes into account both shear and rotary inertia.

The Timoshenko beam theory was published 92 years ago, and since that time it has been successfully used for static and dynamic analysis of beam structural elements, bridges, ship hulls, etc, [1, 2]. In recent decades, it has ordinarily been used for mathematical modeling of very large floating structures in macroscale, like supertankers, ultra-large container ships and floating airports, [3, 4], as well as nanotubes in nanoscale [5, 6].

The Timoshenko beam theory deals with two differential equations of motion with deflection and cross-section rotation as the basic variables. The analytical solution of vibrations in the frequency domain is a rather difficult task and therefore numerical integration of the governing differential equations in the time domain is usually undertaken, limiting in such a way a detailed physical insight.

One of the interesting problems is the analysis of the dynamic response of railway and highway bridges to the moving load of rail and road vehicles, respectively, [7–10]. Actually, there are two types of problems, one with moving gravity force and another with moving inertia force. Dynamic analysis of the former problem is relatively simple and has been treated in numerous articles, mainly in an analytical way by employing the Euler-Bernoulli beam theory, [11-13], and the latter in a semi-analytical way, utilizing the Timoshenko beam theory, [14-16]. The problem of beam vibrations due to moving inertia force is associated with serious difficulties since that force depends on deflection acceleration, [17, 18]. Formulations of the vibration problem of a beam exposed to the action of moving gravity and inertia force and known solutions, with an extensive list of relevant references up to 2012, can be found in a very useful book, [19].

Lumped gravity and inertia force are incorporated into the differential equation of vibrations by the Dirac delta function that is doubtful due to both discontinuity and motion. This can only be denoted as an indication for appropriate treatment when the problem is mapped from the 3D space to the multidimensional space of beam natural modes as new coordinate functions. Therefore, the response of a beam is ordinarily determined by applying the superposition method, utilizing separation of the space and time variables, and the Galerkin method, [5, 6, 16, 17]. The system of modal differential equations is solved by numerical integration in the time domain. Usually, an extensive parametric study is performed, the influence of particular parameters on response is registered, but a detailed physical explanation of beam dynamic behavior is not possible.

Noticeable progress in the analysis of dynamic response of a simply supported beam exposed to moving inertia force is presented in [20]. The Timoshenko beam theory and the energy approach are used, and the modal differential equations are derived by employing the Lagrange equation of the second kind. In this direct problem formulation, the application of partial differential equations with doubtful modeling of discontinuity and moving inertia force is avoided.

2 State of the art

There are two approaches to solve the problem of Timoshenko beam vibrations excited by moving gravity and inertia force. The first starts with a differential equation of motion and employs the modal superposition method, the separation of space and time variables, and the Galerkin method. The second approach is based on the energy balance, modal superposition and the separation of variables. The first formulation, commonly used for beam-like macro structures and nanotubes, [16, 18] and [6] respectively, can be summarized as follows. The equilibrium of transverse forces and bending moments in terms of displacements reads

$$k_S GA\left(\frac{\partial^2 w}{\partial x^2} + \frac{\partial \psi}{\partial x}\right) - \rho A \frac{\partial^2 w}{\partial t^2} = -M\left(g - \frac{D^2 w}{Dt^2}\right)\tilde{\delta}(x - x_M), \qquad (1)$$

$$EI\frac{\partial^2 \psi}{\partial x^2} - k_S GA\left(\frac{\partial w}{\partial x} + \psi\right) - \rho I \frac{\partial^2 \psi}{\partial t^2} = 0, \qquad (2)$$

where w and ψ are deflection and cross-section rotation, respectively, A is cross-section area, I is its moment of inertia, E is Young's modulus, $G = E/(2(1+v))$ is shear modulus, k_S is the shear coefficient, ρ is mass density, g is the gravity constant, M is moving mass with constant velocity v and coordinate $x_M = vt$, $D(.)/Dt$ is material derivative and $\tilde{\delta}$ is the Dirac delta function.

Material acceleration is related to the point of moving mass with respect to the fixed coordinate system, and is presented by the Renaudot formula, [19],

$$\frac{D^2 w(x_M,t)}{Dt^2} = \frac{\partial^2 w(x,t)}{\partial t^2}\bigg|_{x_M} + 2v\frac{\partial^2 w(x,t)}{\partial x \partial t}\bigg|_{x_M} + v^2\frac{\partial^2 w(x,t)}{\partial x^2}\bigg|_{x_M}. \qquad (3)$$

5

Three terms on the right-hand side of Eq. (3) represent vertical, Coriolis and centripetal accelerations of the moving mass, respectively. Coupled differential equations (1) and (2) cannot be solved in the usual way by splitting the beam into two parts, due to moving discontinuity. The problem is overcome by employing the Galerkin method.

If a simply supported beam is considered, deflection and cross-section rotation are expressed by a series of products of natural modes and unknown time functions

$$w(x,t) = \sum_{i=1}^{\infty} W_i(x) R_i(t), \quad \psi(x,t) = \sum_{i=1}^{\infty} \Psi_i(x) \Phi_i(t). \tag{4}$$

According to the Galerkin method, Eqs. (1) and (2) are multiplied with mode functions $W_j(x)$ and $\Psi_j(x)$, respectively, and then integrated along the beam. As a result, a matrix modal differential equation is obtained which can be written in the form

$$\begin{bmatrix} \left[\tilde{M}(t)\right]_{11} & [0] \\ [0] & [M]_{22} \end{bmatrix} \begin{Bmatrix} \{\ddot{R}\} \\ \{\ddot{\Phi}\} \end{Bmatrix} + \begin{bmatrix} \left[c(t)\right]_{11} & [0] \\ [0] & [0] \end{bmatrix} \begin{Bmatrix} \{\dot{R}\} \\ \{\dot{\Phi}\} \end{Bmatrix}$$
$$+ \begin{bmatrix} \left[\tilde{K}(t)\right]_{11} & [K]_{12} \\ [K]_{21} & [K]_{22} \end{bmatrix} \begin{Bmatrix} \{R\} \\ \{\Phi\} \end{Bmatrix} = \begin{Bmatrix} \{F(t)\} \\ \{0\} \end{Bmatrix}. \tag{5}$$

By taking the general integration property of the Dirac delta function into account, the material acceleration (3) generates time dependant moving mass inertia, damping and stiffness matrices, $\left[m(t)\right]_{11}$, $\left[v(t)\right]_{11}$ and $\left[k(t)\right]_{11}$, respectively, which are incorporated in (5), i.e. $\left[\tilde{M}(t)\right]_{11} = [M]_{11} + \left[m(t)\right]_{11}$ and $\left[\tilde{K}(t)\right]_{11} = [K]_{11} - \left[k(t)\right]_{11}$. Moving gravity force is transformed into modal external excitation $\{F(t)\}$. Coupling between the flexural and rotational time function is realized by the stiffness matrix. The differential equation (5) is rather complex due to time varying coefficients and therefore it is ordinarily solved in the time domain.

6

A semi-analytical solution of the vibration of a simply supported Timoshenko beam, based on the energy approach, is presented in [20]. The beam kinetic and strain energy is specified as

$$E_K = \frac{1}{2}\rho A \int_0^l \left(\frac{\partial w}{\partial t}\right)^2 dx + \frac{1}{2}\rho I \int_0^l \left(\frac{\partial \psi}{\partial t}\right)^2 dx, \tag{6}$$

$$E_P = \frac{1}{2}EI \int_0^l \left(\frac{\partial \psi}{\partial x}\right)^2 dx + \frac{1}{2}k_s GA \int_0^l \left(\frac{\partial w}{\partial x} + \psi\right)^2 dx. \tag{7}$$

The kinetic energy of moving inertia force and the potential energy of moving gravity force, $P = Mg$, due to vertical vibrations, read

$$E_{MK} + E_{MP} = \frac{1}{2}M \left[\frac{Dw(x_M,t)}{Dt}\right]^2 + P\, w(x_M,t), \tag{8}$$

where the first term includes the material velocity. The deflection and rotation angle are assumed in the form (4) and substituted into (6), (7) and (8). Then the Lagrange equation of the second kind is used to derive coupled modal equations for $R_i(t)$ and $\Phi_i(t)$. Furthermore, the time function $\Phi_i(t)$ is eliminated and the modal equations are presented in the normalized form

$$
\begin{aligned}
&\ddot{R}_i + \beta \sum_{j=1}^{\infty} f_{ij}^{(1)}(t)\ddot{R}_j + 2\beta \sum_{j=1}^{\infty}\left[\omega_i f_{ij}^{(2)}(t) + 2\omega_j f_{ij}^{(3)}(t)\right]\ddot{R}_j \\
&+ \left[\frac{A}{I}c_1^2 + \frac{\omega_i^2}{v^2}\left(c_1^2 + c_2^2\right)\right]\ddot{R}_i + \sum_{j=1}^{\infty}\left\{g_i f_{ij}^{(1)}(t) + 6\beta\left[\omega_i\omega_j f_{ij}^{(4)}(t) - \omega_j^2 f_{ij}^{(3)}(t)\right]\right\}\dot{R}_j \\
&+ 2\sum_{j=1}^{\infty}\left\{g_i\omega_j f_{ij}^{(3)}(t) - \beta\left[3\omega_i\omega_j^2 f_{ij}^{(2)}(t) + 2\omega_j^3 f_{ij}^{(3)}(t)\right]\right\}\dot{R}_j \\
&+ \frac{\omega_i^4}{v^4}c_1^2 c_2^2 R_i - \sum_{j=1}^{\infty}\left\{g_i\omega_j^2 f_{ij}^{(1)}(t) + \beta\left[2\omega_i\omega_j^3 f_{ij}^{(4)}(t) - \omega_j^4 f_{ij}^{(1)}(t)\right]\right\}R_j = \frac{P}{\rho A \beta}g_i \sin\omega_i t,
\end{aligned} \tag{9}
$$

where

$$c_1 = \sqrt{\frac{k_s G}{\rho}}, \; c_2 = \sqrt{\frac{E}{\rho}}, \; \beta = \frac{2M}{\rho A l}, \; \omega_i = \frac{i\pi v}{l}, \; \omega_j = \frac{j\pi v}{l}, \tag{10}$$

$$g_i = \beta \left[\frac{A}{I} c_1^2 + \omega_i^2 \left(\frac{c_2^2}{v^2} - 1 \right) \right], \tag{11}$$

and

$$f_{ij}^{(1)}(t) = \sin \omega_i t \sin \omega_j t, \; f_{ij}^{(2)}(t) = \cos \omega_i t \sin \omega_j t,$$
$$f_{ij}^{(3)}(t) = \sin \omega_i t \cos \omega_j t, \; f_{ij}^{(4)}(t) = \cos \omega_i t \cos \omega_j t. \tag{12}$$

The system of differential equations (9) is rather complex due to variable coefficients and is solved numerically in [20].

3 Motivation and research outline

Investigation of this challenging problem, in spite of a very large number of published papers, is motivated by the state of the art. It seems that all available possibilities have not yet been used. The problem of Timoshenko beam vibrations excited by moving gravity and inertia force can be formulated in different ways, and it is reasonable to use the simplest one. Governing modal differential equations have not yet been completely solved analytically, as in the case of the Euler-Bernoulli beam, [17]. Furthermore, the influence of damping on response has not been analyzed in detail.

In such circumstances, different mathematical models for Timoshenko beam vibration analyses are presented in this book. Two 2nd order partial differential equations with distributed stationary dynamic load are transformed into a single partial differential equation of the 4th order, once in terms of deflection and then in terms of rotation angle. Modal equations are derived by employing the Galerkin method and the energy balance for both deflection and rotation angle. As a result, six different formulations of load function are obtained and compared.

Since the application of the original Timoshenko beam theory is rather complex in some cases, it is modified in such a way that total deflection and rotation angle are decomposed into bending and shear deflection, and bending cross-section rotation and in-plane shear slope angle, respectively, [21]. In this way, a single 4th order partial differential equation of flexural vibrations in terms of bending deflection and a 2nd order partial differential equation for in-plane shear vibrations are obtained. Modal equations of flexural vibrations are derived by using the Galerkin method and the energy approach. The obtained functions of distributed dynamic load are compared with those determined by the original Timoshenko beam theory. It is found that the modified Timoshenko beam theory in combination with the energy balance is the simplest one, and therefore it is used in the subsequent dynamic analysis.

A simply supported beam is considered and a distributed dynamic load is lumped and transformed into moving gravity and inertia force. Gravity force causes linear harmonic vibrations, while moving inertia force excites parametric vibrations which are solved by the perturbation method. Hence, linear and parametric vibrations are analyzed in the frequency domain with damping included. Parametric excitation is expanded into harmonics and the harmonic balance method is used to achieve an analytical solution. Special attention is paid to resonant parametric vibrations.

4 Application of the Timoshenko beam theory

4.1 Differential equation of vibrations

Derivation of the Timoshenko beam theory is outlined in order to achieve a proper formulation of the excitation function. The Timoshenko beam theory operates with two variables, i.e. beam deflection and rotation of cross-section, w and ψ, respectively, [1, 2]. The bending moment and shear force read

$$M = D\frac{\partial \psi}{\partial x}, \quad Q = S\left(\frac{\partial w}{\partial x} + \psi\right), \tag{13}$$

where $D = EI$ is flexural rigidity and $S = k_s GA$ is shear rigidity. The stiffness parameters for complex thin-walled girders, like ship hulls, are determined by the strip element method, [3].

The beam is loaded with an external distributed dynamic load, $q(x,t)$, and the inertia load and bending moment are

$$q_I = -m\frac{\partial^2 w}{\partial t^2}, \quad m_I = -J\frac{\partial^2 \psi}{\partial t^2}, \tag{14}$$

where $m = \rho A$ is the specific mass per unit length, $J = \rho I$ is the mass moment of inertia and ρ is mass density.

The equilibrium of moments and forces,

$$\frac{\partial M}{\partial x} - Q = -m_I, \quad \frac{\partial Q}{\partial x} = -q_I - q, \tag{15}$$

leads to a system of coupled second order partial differential equations of motion

11

$$D\frac{\partial^2\psi}{\partial x^2} - S\left(\frac{\partial w}{\partial x}+\psi\right) - J\frac{\partial^2\psi}{\partial t^2} = 0, \tag{16}$$

$$S\left(\frac{\partial^2 w}{\partial x^2}+\frac{\partial\psi}{\partial x}\right) - m\frac{\partial^2 w}{\partial t^2} = -q. \tag{17}$$

Eqs. (16) and (17) are widely used in the literature for beam vibrations analysis. However, handling is not so simple in some cases, and therefore Eqs. (16) and (17) are condensed into a single one of the fourth order. From (17) one obtains

$$\frac{\partial\psi}{\partial x} = -\frac{\partial^2\psi}{\partial x^2} + \frac{m}{S}\frac{\partial^2 w}{\partial t^2} - \frac{q}{S}, \tag{18}$$

and by substituting (18) into (16) derived per x, one obtains

$$\frac{\partial^4 w}{\partial x^4} - \left(\frac{m}{S}+\frac{J}{D}\right)\frac{\partial^4 w}{\partial x^2\partial t^2} + \frac{m}{D}\frac{\partial^2 w}{\partial t^2} + \frac{mJ}{DS}\frac{\partial^4 w}{\partial t^4} = \frac{1}{D}\left(q - \frac{D}{S}\frac{\partial^2 q}{\partial x^2} + \frac{J}{S}\frac{\partial^2 q}{\partial t^2}\right). \tag{19}$$

Once Eq. (19) is solved, the angle of rotation of the cross-section is determined by (6) as

$$\psi = -\frac{\partial w}{\partial x} + \frac{m}{S}\int\frac{\partial^2 w}{\partial t^2}dx + f(t), \tag{20}$$

where $f(t)$ is a time function of rigid body motion. Sectional forces are obtained by Eqs. (13).

Another possibility is to extract deflection from Eq. (16), i.e.

$$\frac{\partial w}{\partial x} = \frac{D}{S}\frac{\partial^2\psi}{\partial x^2} - \frac{J}{S}\frac{\partial^2\psi}{\partial t^2} - \psi, \tag{21}$$

and to substitute it into (17). In that case, one obtains

$$\frac{\partial^4 \psi}{\partial x^4} - \left(\frac{m}{S} + \frac{J}{D}\right)\frac{\partial^4 \psi}{\partial x^2 \partial t^2} + \frac{m}{D}\frac{\partial^2 \psi}{\partial t^2} + \frac{mJ}{DS}\frac{\partial^4 \psi}{\partial t^4} = -\frac{1}{D}\frac{\partial q}{\partial x}. \tag{22}$$

Once Eq. (22) is solved, deflection is obtained from (21)

$$w = \frac{D}{S}\frac{\partial \psi}{\partial x} - \frac{J}{S}\int \frac{\partial^2 \psi}{\partial t^2}dx - \int \psi dx + g(t), \tag{23}$$

where $g(t)$ is a time-dependent function of rigid body motion.

The differential operator in both Eq. (19) and (22) is the same, while the difference is in the load function, which is simpler in the latter case. However, deflection w is obtained directly from (19), and indirectly from (22).

It is necessary to point out that the system of two coupled 2nd order equilibrium equations (1) and (2), with a priori specified moving gravity and inertia force, can also be transformed into two 4th order uncoupled equations like (19) and (22). However, the excitation function takes a rather complex form compared to the ones in (19) and (22), which complicates further handling.

4.2 Modal differential equations formulated by the Galerkin method

Since the beam is exposed to a moving load, the governing partial differential equation of motion (19) or (22) cannot be solved directly. Therefore, the modal superposition method, the separation of space and time variables and the Galerkin method are used. Accordingly, the beam deflection is assumed in the form (4) which is substituted into (19), multiplied by the mode function $W_i(x)$ and integrated along the beam. As a result, a system of modal differential equations is obtained, which is written in the matrix notation

13

$$\left[N_{ij} \right]\{\dddot{R}_j\} + \left[M_{ij} \right]\{\ddot{R}_j\} + \left[K_{ij} \right]\{R_j\} = \{F_{TGR}^i\}, \tag{24}$$

where

$$N_{ij} = \frac{mJ}{DS} I_{ij}^{(0)}, \ M_{ij} = \frac{m}{D} I_{ij}^{(0)} - \left(\frac{m}{S} + \frac{J}{D} \right) I_{ij}^{(2)}, \ K_{ij} = I_{ij}^{(4)}, \tag{25}$$

$$I_{ij}^{(0)} = \int_0^l W_i W_j \mathrm{d}x, \ I_{ij}^{(2)} = \int_0^l W_i W_j'' \mathrm{d}x, \ I_{ij}^{(4)} = \int_0^l W_i W_j'''' \mathrm{d}x, \tag{26}$$

and

$$F_{TGR}^i = \frac{1}{D} \int_0^l \left(q - \frac{D}{S} \frac{\partial^2 q}{\partial x^2} + \frac{J}{S} \frac{\partial^2 q}{\partial t^2} \right) W_i \mathrm{d}x \tag{27}$$

are generalized force, and l is beam length.

If the partial differential equation (22) is taken into account and the rotation angle is assumed in the form (4), the corresponding modal differential equations are

$$\left[N_{ij} \right]\{\dddot{\Phi}_j\} + \left[M_{ij} \right]\{\ddot{\Phi}_j\} + \left[K_{ij} \right]\{\Phi_j\} = \{F_{TG\Phi}^i\}, \tag{28}$$

where the matrices are of the same form as (25) and

$$I_{ij}^{(0)} = \int_0^l \Psi_i \Psi_j \mathrm{d}x, \ I_{ij}^{(2)} = \int_0^l \Psi_i \Psi_j'' \mathrm{d}x, \ I_{ij}^{(4)} = \int_0^l \Psi_i \Psi_j'''' \mathrm{d}x, \tag{29}$$

while

$$F_{TG\Phi}^i = -\frac{1}{D}\int_0^l \frac{\partial q}{\partial x}\Psi_i dx. \tag{30}$$

Once functions $\Phi_j(t)$ are determined, the deflection is obtained by Eq. (23), i.e.

$$w = -\sum_{j=0}^{\infty}\left\{\left[\int\Psi_j dx - \frac{D}{S}\Psi_j'\right]\Phi_j + \frac{J}{S}\int\Psi_j dx\ddot{\Phi}_j\right\} + g(t). \tag{31}$$

4.3 Modal differential equations formulated by energy balance

The total energy of a vibrating beam consists of kinetic energy, E_k, strain energy, E_p, and the work of the external dynamic load $V = \int_0^l qw dx$. The kinetic energy includes transverse and rotary inertia terms, while strain energy due to the bending moment and shear force is

$$E_k = \frac{1}{2}m\int_0^l\left(\frac{\partial w}{\partial t}\right)^2 dx + \frac{1}{2}J\int_0^l\left(\frac{\partial \psi}{\partial t}\right)^2 dx, \tag{32}$$

$$E_p = \frac{1}{2}D\int_0^l\left(\frac{\partial \psi}{\partial x}\right)^2 dx + \frac{1}{2}S\int_0^l\left(\frac{\partial w}{\partial x}+\psi\right)^2 dx. \tag{33}$$

The deflection and rotation angle are specified by Eqs. (4), respectively. Beam energy has to satisfy the Lagrange equation of the second kind, once per R_i and another time per Φ_i, i.e.

$$\frac{d}{dt}\left(\frac{\partial E_k}{\partial \dot{R}_i}\right) - \frac{\partial E_k}{\partial R_i} + \frac{\partial E_p}{\partial R_i} = \frac{\partial V}{\partial R_i}, \tag{34}$$

$$\frac{\mathrm{d}}{\mathrm{d}t}\left(\frac{\partial E_k}{\partial \dot{\Phi}_i}\right) - \frac{\partial E_k}{\partial \Phi_i} + \frac{\partial E_p}{\partial \Phi_i} = \frac{\partial V}{\partial \Phi_i}. \tag{35}$$

The general formulation of modal equations for a beam with arbitrary boundary conditions is a rather difficult task and therefore a simply supported beam is considered. Natural modes are assumed in the form $W_j = \sin(j\pi x/l)$ and $\Psi_j = \cos(j\pi x/l)$. Since they are orthogonal, the integrals occurring in Eqs. (34) and (35) are the following

$$\int_0^l W_i^2 \mathrm{d}x = \int_0^l \Psi_i^2 \mathrm{d}x = \frac{l}{2}, \quad \int_0^l W_i' \, \Psi_i \mathrm{d}x = \int_0^l W_i \Psi_i' \mathrm{d}x = \frac{i\pi}{l}\frac{l}{2},$$
$$\int_0^l \left(W_i'\right)^2 \mathrm{d}x = \int_0^l \left(\Psi_i'\right)^2 \mathrm{d}x = \left(\frac{i\pi}{l}\right)^2 \frac{l}{2}. \tag{36}$$

As a result, two coupled modal differential equations are obtained

$$m\frac{l}{2}\ddot{R}_i + S\left(\frac{i\pi}{l}\right)^2 \frac{l}{2} R_i + S\left(\frac{i\pi}{l}\right)\frac{l}{2}\Phi_i = \int_0^l q W_i \mathrm{d}x, \tag{37}$$

$$J\frac{l}{2}\ddot{\Phi}_i + S\left[1+\left(\frac{i\pi}{l}\right)^2 \frac{D}{S}\right]\frac{l}{2}\Phi_i + S\left(\frac{i\pi}{l}\right)\frac{l}{2}R_i = 0. \tag{38}$$

By taking Φ_i from (37) and substituting it into (38), one arrives at a single modal differential equation of the 4th order

$$\frac{mJ}{DS}\frac{l}{2}\ddddot{R}_i + \left[\frac{m}{D}+\left(\frac{i\pi}{l}\right)^2\left(\frac{m}{S}+\frac{J}{D}\right)\right]\frac{l}{2}\ddot{R}_i + \left(\frac{i\pi}{l}\right)^4\frac{l}{2}R_i = F_{TER}^i, \tag{39}$$

where

16

$$F_{TER}^i = \frac{1}{D} \int_0^l \left\{ \left[1 + \left(\frac{i\pi}{l} \right)^2 \frac{D}{S} \right] qW_i + \frac{J}{S} \frac{\partial^2}{\partial t^2} (qW_i) \right\} dx. \tag{40}$$

If function R_i is extracted from (38) and substituted into (37), one obtains

$$\frac{mJ}{DS} \frac{l}{2} \dddot{\Phi}_i + \left[\frac{m}{D} + \left(\frac{i\pi}{l} \right)^2 \left(\frac{m}{S} + \frac{J}{D} \right) \right] \frac{l}{2} \ddot{\Phi}_i + \left(\frac{i\pi}{l} \right)^4 \frac{l}{2} \Phi_i = F_{TE\Phi}^i, \tag{41}$$

where

$$F_{TE\Phi}^i = -\frac{1}{D} \left(\frac{i\pi}{l} \right) \int_0^l qW_i dx. \tag{42}$$

Once Φ_i is determined, R_i is obtained from (38), and total deflection according to (4) reads

$$w = -\sum_{i=0}^{\infty} \left(\frac{l}{i\pi} \right) \sin \frac{i\pi x}{l} \left\{ \left[1 + \left(\frac{i\pi}{l} \right)^2 \frac{D}{S} \right] \Phi_i + \frac{J}{S} \ddot{\Phi}_i \right\}. \tag{43}$$

5 Application of the modified Timoshenko beam theory

5.1 Differential equation of vibrations

The Timoshenko beam theory is modified in such a way that the beam deflection and angle of rotation are split into their constitutive parts

$$w = w_b + w_s, \ \psi = \varphi + \vartheta, \ \varphi = -\frac{\partial w_b}{\partial x}, \tag{44}$$

where w_b and w_s are beam deflection due to bending and shear, respectively, and φ is the angle of cross-section rotation due to bending, while ϑ is the cross-section slope angle due to axial shear (like the sliding of a deck of playing cards), [21]. Substituting (44) into Eqs. (16) and (17) yields

$$D\frac{\partial^3 w_b}{\partial x^3} - J\frac{\partial^2}{\partial t^2}\left(\frac{\partial w_b}{\partial x}\right) + S\frac{\partial w_s}{\partial x} = D\frac{\partial^2 \vartheta}{\partial x^2} - S\vartheta - J\frac{\partial^2 \vartheta}{\partial t^2}, \tag{45}$$

$$S\frac{\partial^2 w_s}{\partial x^2} - m\frac{\partial^2}{\partial t^2}\left(w_b + w_s\right) = -S\frac{\partial \vartheta}{\partial x} - q. \tag{46}$$

Since three variables are present in (45) and (46) and only two equations are on disposal, it is reasonable to assume that the flexural and in-plane shear displacement fields are not coupled, [21]. In that case, by setting both the left and right hand side of Eq. (45) equal to zero, one obtains from the former

$$w_s = -\frac{D}{S}\frac{\partial^2 w_b}{\partial x^2} + \frac{J}{S}\frac{\partial^2 w_b}{\partial t^2}. \tag{47}$$

By substituting (47) into (46), a partial differential equation of the 4th order for flexural vibrations in terms of bending deflection is obtained

$$\frac{\partial^4 w_b}{\partial x^4} - \left(\frac{m}{S} + \frac{J}{D}\right)\frac{\partial^4 w_b}{\partial x^2 \partial t^2} + \frac{m}{D}\frac{\partial^2 w_b}{\partial t^2} + \frac{mJ}{DS}\frac{\partial^4 w_b}{\partial t^4} = \frac{q}{D}. \tag{48}$$

Once w_b is determined, the total beam deflection, according to (44) and (47), reads

$$w = w_b - \frac{D}{S}\frac{\partial^2 w_b}{\partial x^2} + \frac{J}{S}\frac{\partial^2 w_b}{\partial t^2}, \tag{49}$$

while in-plane shear, described by the separated terms in Eq. (45),

$$\frac{\partial^2 \vartheta}{\partial x^2} - \frac{S}{D}\vartheta - \frac{J}{D}\frac{\partial^2 \vartheta}{\partial t^2} = 0, \tag{50}$$

is analyzed in [21] as a specific task.

5.2 Modal differential equations formulated by the Galerkin method

Following the procedure described in Section 2.2, beam bending deflection is assumed in the form

$$w_b(x,t) = \sum_{j=1}^{\infty} W_j(x)T_j(t), \tag{51}$$

and substituted into (48), which is further multiplied by W_i and integrated along the beam. As a result, a modal differential equation of motion is obtained

$$\left[N_{ij}\right]\{\ddddot{T}_j\} + \left[M_{ij}\right]\{\ddot{T}_j\} + \left[K_{ij}\right]\{T_j\} = \{F_{MTG}^i\}, \tag{52}$$

19

where matrices and integrals are of the same form as Eqs. (25) and (26), while the element of the modal load vector reads

$$F_{MTG}^i = \frac{1}{D}\int_0^l qW_i dx.$$ (53)

The total deflection is obtained from (49), i.e.

$$w = \sum_{j=1}^{\infty}\left\{\left[W_j - \frac{D}{S}W_j''\right]T_j + \frac{J}{S}W_j\ddot{T}_j\right\}.$$ (54)

5.3 Modal differential equations formulated by energy balance

In the considered case, bending deflection and total deflection are actual. The corresponding kinetic and strain energy as well as the work of external dynamic force read

$$E_k = \frac{1}{2}m\int_0^l\left(\frac{\partial w}{\partial t}\right)^2 dx + \frac{1}{2}J\int_0^l\left(\frac{\partial^2 w_b}{\partial x\partial t}\right)^2 dx,$$ (55)

$$E_p = \frac{1}{2}D\int_0^l\left(\frac{\partial^2 w}{\partial x^2}\right)^2 dx + \frac{1}{2}S\int_0^l\left(\frac{\partial w}{\partial x} - \frac{\partial w_b}{\partial x}\right)^2 dx, \quad V = \int_0^l qw dx.$$ (56)

Total and bending deflection is assumed by the modal series (4) and (51), respectively. The simply supported beam is analyzed using $W_j = \sin(j\pi x/l)$. The Lagrange equation of the second kind is employed as in Section 2.3, Eqs. (34) and (35), with derivatives per time functions R_i and T_i. By substituting (4) and (51) into the Lagrange equations, the following integrals occur

20

$$\int_0^l W_i^2 dx = \frac{l}{2}, \quad \int_0^l (W_i')^2 \, dx = \left(\frac{i\pi}{l}\right)^2 \frac{l}{2}, \quad \int_0^l (W_i'')^2 \, dx = \left(\frac{i\pi}{l}\right)^4 \frac{l}{2}. \tag{57}$$

As a result, two coupled modal differential equation are obtained

$$m\frac{l}{2}\ddot{R}_i + S\left(\frac{i\pi}{l}\right)^2 \frac{l}{2}(R_i - T_i) = \int_0^l q W_i dx, \tag{58}$$

$$J\left(\frac{i\pi}{l}\right)^2 \frac{l}{2}\ddot{T}_i + S\left(\frac{i\pi}{l}\right)^2 \left[1 + \left(\frac{i\pi}{l}\right)^2 \frac{D}{S}\right]\frac{l}{2}T_i - S\left(\frac{i\pi}{l}\right)^2 \frac{l}{2}R_i = 0. \tag{59}$$

By taking R_i from (59) and substituting it into (58), a single modal differential equation is obtained

$$\frac{mJ}{DS}\frac{l}{2}\dddot{T}_i + \left[\frac{m}{D} + \left(\frac{i\pi}{l}\right)^2 \left(\frac{m}{S} + \frac{J}{D}\right)\right]\frac{l}{2}\ddot{T}_i + \left(\frac{i\pi}{l}\right)^4 \frac{l}{2}T_i = F_{MTE}^i, \tag{60}$$

where

$$F_{MTE}^i = \frac{1}{D}\int_0^l q W_i dx. \tag{61}$$

Once function T_i is known, R_i is determined from (59), and total deflection, according to (4), reads

$$w = \sum_{i=1}^{\infty} \sin\frac{i\pi x}{l} \left\{\left[1 + \left(\frac{i\pi}{l}\right)^2 \frac{D}{S}\right]T_i + \frac{J}{S}\ddot{T}_i\right\}. \tag{62}$$

21

If function T_i is eliminated from the system of Eqs. (62) and (63), one obtains a modal differential equation in terms of the total deflection time function R_i, which is identical to Eq. (41) derived within direct application of the Timoshenko beam theory.

6 Comparison of different formulations of differential equations

6.1 Application of the Timoshenko beam theory

A single partial differential equation of vibrations in terms of total deflection and rotation is derived, Eqs. (19) and (22), respectively. The former form was obtained by Timoshenko himself for natural vibration analysis, [22], but it is very seldom used in the relevant literature. Eq. (22) is new and its advantage is a simpler loading function and its disadvantage is the indirect determination of the total deflection by Eq. (23).

Transformation of the partial differential equation of vibrations into ordinary modal differential equations by the Galerkin method is also given in two alternatives, i.e. in terms of the total deflection time function, (24), and in terms of the rotary angle time function, (28), with the same characteristics as in the previous case. Both formulations are applicable for a beam with arbitrary boundary conditions.

The formulation of modal differential equations by energy balance is a somewhat difficult task due to the application of the Lagrange equation of the second kind. Therefore, in this case the modal equations are derived only for the case of a simply supported beam also in two alternatives, as in the previous case, Eqs. (39) and (41). The direct determination of the deflection time function R_i by solving Eq. (39) seems to be preferable. However, Eq. (39) includes a more complicated excitation function F_{TER}^i, Eq. (40), than Eq. (28) for Φ_i, i.e. $F_{TE\Phi}^i$, Eq. (42). Also, Φ_i extracted from (37) depends on q directly and indirectly through R_i, which is not practical for the determination of sectional forces, Eqs. (13).

Different partial differential equations of vibrations and different ordinary modal differential equations, derived in Section 2, have the same differential operator (left hand side), respectively. Their difference is manifested in the loading function (right hand side), as can be seen in Table 1.

There is a small difference in the last term between loading functions derived in terms of total deflection by the Galerkin method and energy approach, Eqs. (27) and (40), respectively, Table 1. Namely, the former contains $\dfrac{\partial^2 q}{\partial t^2} W_i$, while the later has $\dfrac{\partial^2}{\partial t^2}(q W_i)$. This difference does not have any influence if a stationary dynamic load is under consideration, but has repercussions on the accuracy of the Galerkin formulation in the case of a moving load, which will be discussed later. Such a difference does not occur if modal equations are formulated in terms of rotary angle time functions, Eqs. (30) and (42), Table 1.

Table 1 Loading function in the Timoshenko beam theory

	Total deflection, w	Rotation angle, ψ
Differential equation	$\dfrac{1}{D}\left(q - \dfrac{D}{S}\dfrac{\partial^2 q}{\partial x^2} + \dfrac{J}{S}\dfrac{\partial^2 q}{\partial t^2}\right)$, Eq. (19)	$-\dfrac{1}{D}\dfrac{\partial q}{\partial x}$ Eq., (22)
Modal equations, Galerkin method	$\dfrac{1}{D}\displaystyle\int_0^l \left(q - \dfrac{D}{S}\dfrac{\partial^2 q}{\partial x^2} + \dfrac{J}{S}\dfrac{\partial^2 q}{\partial t^2}\right)W_i dx$, Eq. (27)	$-\dfrac{1}{D}\displaystyle\int_0^l \dfrac{\partial q}{\partial x}\Phi_i dx$, Eq. (30)
Modal equations, energy balance	$\dfrac{1}{D}\displaystyle\int_0^l \left\{\left[1+\left(\dfrac{i\pi}{l}\right)^2 \dfrac{D}{S}\right]q W_i + \dfrac{J}{S}\dfrac{\partial^2}{\partial t^2}(q W_i)\right\}dx$, Eq. (40)	$-\dfrac{1}{D}\left(\dfrac{i\pi}{l}\right)\displaystyle\int_0^l q W_i dx$, Eq. (42)

6.2 Application of the modified Timoshenko beam theory

A partial differential equation of vibrations in terms of bending deflection (48) is of the same structure as (19) derived in terms of total deflection by applying the Timoshenko beam theory, but with a simpler loading function, Table 1 and 2, respectively. Therefore, it is easier to solve Eq. (48) and determine total deflection indirectly by Eq. (49) than to undertake integration of Eq. (19) for the direct determination of total deflection. It is interesting to note that the excitation function in (19) has the same differential operator as function (49) for total deflection. Also, it is important to point out that modal differential equations derived by the Galerkin method

24

and energy balance are identical, Eqs. (52) and (60), respectively, including their loading functions, Eqs. (53) and (61), Table 1. However, this is not the case if the Timoshenko beam theory is used, Eqs. (27) and (40).

Based on the above facts and more complex modal loading function, Eq. (40), compared to (53) or (61), modal differential equations obtained by the modified Timoshenko beam theory in combination with the Galerkin method or energy balance, Eqs. (52) and (60), are considered further on.

Table 2 Loading function in the modified Timoshenko beam theory

	Total deflection, w	Bending deflection, w_b
Differential equation		$\dfrac{q}{D}$ Eq., (48)
Modal equations, Galerkin method		$\dfrac{1}{D}\displaystyle\int_0^l qW_i dx$, Eq. (53)
Modal equations, energy balance	$\dfrac{1}{D}\displaystyle\int_0^l \left\{ \left[1 + \left(\dfrac{i\pi}{l} \right)^2 \dfrac{D}{S} \right] qW_i + \dfrac{J}{S}\dfrac{\partial^2}{\partial t^2}(qW_i) \right\} dx$, Eq. (40)	$\dfrac{1}{D}\displaystyle\int_0^l qW_i dx$, Eq. (61)

25

7 Forced vibrations excited by moving gravity force

A gravity force P is moving along the beam with constant velocity v_p, so that its coordinate is $x = v_p t$, Fig. 1. In that case, the mode function is $W_i = \sin(i\pi x/l) = \sin\Omega_i t$, where $\Omega_i = i\pi v_p/l$ is excitation frequency. The distributed load in (61) can be presented as $q = P/dx$, and one obtains for the modal load

$$F_i = \frac{P}{D}\sin\Omega_i t. \tag{63}$$

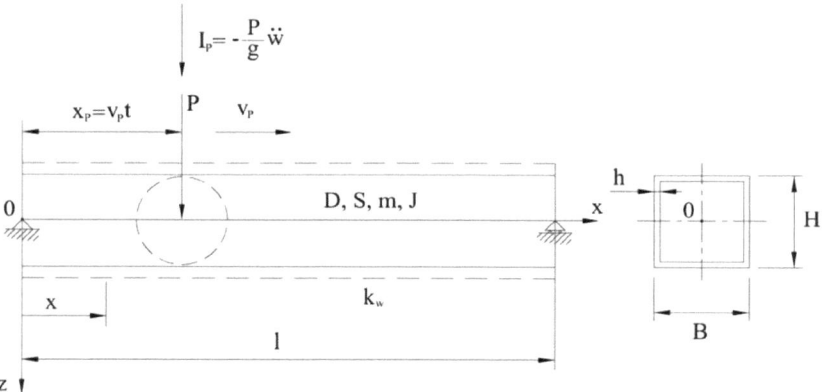

Fig. 1. Timoshenko beam exposed to moving gravity and inertia force

In the literature, the Dirac delta function is used for the transition of a distributed load to a lumped one, with the same result. The modal differential equation (60) is extended with damping force and is presented in the form

$$N_i\dddot{T}_i + M_i\ddot{T}_i + V_i\dot{T}_i + K_i T_i = F_0\sin\Omega_i t,\ F_0 = \frac{P}{D}, \tag{64}$$

where

$$N_i = \frac{mJ}{DS}\frac{l}{2}, \quad M_i = \left[\frac{m}{D} + \left(\frac{i\pi}{l}\right)^2 \left(\frac{m}{S} + \frac{J}{D}\right)\right]\frac{l}{2}, \quad K_i = \left(\frac{i\pi}{l}\right)^4 \frac{l}{2}. \tag{65}$$

The solution of (64) consists of particular and homogenous parts, $T_i = T_i^P + T_i^h$. The former is assumed in the same harmonic form as the excitation

$$T_i^P = A_i \cos\Omega_i t + B_i \sin\Omega_i t. \tag{66}$$

By substituting (66) into (64) and equalizing the coefficients of the sine and cosine functions, a system of two algebraic equations is obtained. Its solution reads

$$A_i = \frac{D_A^i}{D_0^i}, \quad B_i = \frac{D_B^i}{D_0^i}, \tag{67}$$

where

$$
\begin{aligned}
D_A^i &= -\Omega_i V_i F_0, \\
D_B^i &= \left(\Omega_i^4 N_i - \Omega_i^2 M_i + K_i\right) F_0, \\
D_0^i &= \left(\Omega_i^4 N_i - \Omega_i^2 M_i + K_i\right)^2 + \Omega_i^2 V_i^2.
\end{aligned} \tag{68}
$$

A homogenous solution is determined, approximately omitting the first term in Eq. (64) as a small quantity for reasons of simplicity. Hence, [23]

$$T_i^h = C_i e^{-\gamma_i \omega_i t} \cos\omega_i^* t + D_i e^{-\gamma_i \omega_i t} \sin\omega_i^* t, \quad \omega_i^* = \omega_i \sqrt{1-\gamma^2}. \tag{69}$$

where ω_i and ω_i^* are the natural frequency of a conservative and damped system, respectively, and $\gamma_i = V_i / (2\omega_i M_i)$ is the dimensionless damping coefficient. The natural frequency ω_i is determined from the homogenous part of Eq. (64), omitting damping

$$\omega_i^4 N_i - \omega_i^2 M_i + K_i = 0, \tag{70}$$

that gives

$$\omega_i^{(1,2)} = \frac{1}{\sqrt{2N_i}} \sqrt{M_i \pm \sqrt{M_i^2 - 4N_i K_i}}. \tag{71}$$

Two frequency spectra are obtained as a phenomenon of a simply supported beam, [21] and [24].

At the beginning of the moving gravity force action, the beam is at rest. The total time function $T_i = T_i^p + T_i^h$ has to satisfy the initial conditions $T_i(0) = 0$ and $\dot{T}_i(0) = 0$. Hence, one finds

$$
\begin{aligned}
T_i = A_i & \left[\cos \Omega_i t - e^{-\gamma_i \omega_i t} \left(\cos \omega_i^* t + \gamma_i \frac{\omega_i}{\omega_i^*} \sin \omega_i^* t \right) \right] \\
& + B_i \left[\sin \Omega_i t - \frac{\Omega_i}{\omega_i^*} e^{-\gamma_i \omega_i t} \sin \omega_i^* t \right].
\end{aligned}
\tag{72}
$$

In the case of the resonance of a conservative system, $\Omega_i = \omega_i$ and solution (72) does not satisfy the differential equation (64). Therefore, the solution is assumed in the form

$$T_i = A_{ri} \left(\sin \omega_i t - \omega_i t \cos \omega_i t \right). \tag{73}$$

By substituting (73) into (64) and omitting damping, one finds

$$A_{ri} = \frac{F_0}{-3\omega_i^4 N_i + \omega_i^2 M_i + K_i} = \frac{F_0}{2\omega_i^2 \left(M_i - 2\omega_i^2 N_i \right)}. \qquad (74)$$

Constant A_{ri} is rearranged by employing the frequency equation (70). Function (73) satisfies the initial conditions $T_i(0) = 0$ and $\dot{T_i}(0) = 0$. Since the passing time of gravity force along the beam is $t_p = l/v_P$ and $\Omega_i = i\pi v_P/l = \omega_i$, one finds $\omega_i t_p = i\pi$ and the time function reaches the maximum value at t_p, i.e. $T_{ri}(t_p) = i\pi/A_{ri}$.

8 Forced vibrations excited by moving gravity and inertia force

8.1 Modal inertia forces

Assuming the permanent contact of a moving mass P/g along the beam, i.e. $\ddot{w} \le g$, its inertia force, according to (49) and (50), Fig. 1, reads

$$I_P\big(x_P(t),t\big) = -\frac{P}{g}\ddot{w}\big(x_P(t),t\big) =$$
$$-\frac{P}{g}\sum_{j=1}^{\infty}\left[\frac{\partial^2}{\partial t^2}\big(W_jT_j\big) - \frac{D}{S}\frac{\partial^2}{\partial t^2}\left(\frac{\partial^2 W_j}{\partial x^2}T_j\right) + \frac{J}{S}\frac{\partial^4}{\partial t^4}\big(W_jT_j\big)\right]. \tag{75}$$

Since deflection modes are sinusoidal, $W_j = \sin\big(j\pi x/l\big)$, and the mass coordinate is $x_P = v_P t$, we obtain $W_j = \sin\Omega_j t$, where $\Omega_j = j\pi v_P/l$ is modal forcing frequency. By taking into account that the second term in (75) is zero, one obtains

$$I_P(t) = -\frac{P}{g}\sum_{j=1}^{\infty}\left[\frac{\partial^2}{\partial t^2}\big(\sin\Omega_j t\, T_j\big) + \frac{J}{S}\frac{\partial^4}{\partial t^4}\big(\sin\Omega_j t\, T_j\big)\right]. \tag{76}$$

The derivatives in (76) can be presented as the products of two vectors

$$g_j^{(1)} = \frac{\mathrm{d}^2}{\mathrm{d}t^2}\big(\sin\Omega_j t\, T_j\big) = \big\langle\sin\Omega_j t, 2\Omega_j\cos\Omega_j t, -\Omega_j^2\sin\Omega_j t\big\rangle \begin{Bmatrix} \ddot{T}_j \\ \dot{T}_j \\ T_j \end{Bmatrix}, \tag{77}$$

$$g_j^{(2)} = \frac{\mathrm{d}^4}{\mathrm{d}t^4}\big(\sin\Omega_j t\, T_j\big)$$
$$= \big\langle\sin\Omega_j t, 4\Omega_j\cos\Omega_j t, -6\Omega_j^2\sin\Omega_j t, -4\Omega_j^3\cos\Omega_j t, \Omega_j^4\sin\Omega_j t\big\rangle \tag{78}$$
$$\cdot\big\langle\ddddot{T}_j, \dddot{T}_j, \ddot{T}_j, \dot{T}_j, T_j\big\rangle^{\mathrm{T}},$$

where $\langle . \rangle^{\mathrm{T}} = \{.\}$. Modal force due to gravity force is determined by (63) and in the case of inertia force it is necessary to take $I_p(t)$ instead of P. Hence, one obtains

$$F_i = -\frac{P}{gD} \sum_{j=1}^{\infty} \left(g_j^{(1)} \sin \Omega_i t + \frac{J}{S} g_j^{(2)} \sin \Omega_i t \right). \tag{79}$$

Eq. (79) can be written in the matrix notation, and only the first term is expanded for illustration

$$\{F_i\}^{(1)} = -\frac{P}{gD} \left[\sin \Omega_i t \sin \Omega_j t \right] \{\ddot{T}_j\} - \frac{P}{gD} \left[\Omega_j \sin \Omega_i t \cos \Omega_j t \right] \{\dot{T}_j\}$$
$$+ \frac{P}{gD} \left[\Omega_j^2 \sin \Omega_i t \sin \Omega_j t \right] \{T_j\}. \tag{80}$$

The second term in (79), i.e. $\{F_i\}^{(2)}$, can be presented in a similar manner.

8.2 Modal differential equations

The products of the trigonometric functions in (80) are expanded into harmonics by using trigonometric identities. A constant term is extracted from the variable ones

$$\sin \Omega_i t \sin \Omega_j t = \frac{1}{2} \delta_{ij} + \frac{1}{2} \left(1 - \delta_{ij} \right) \cos \left(\Omega_i - \Omega_j \right) t - \frac{1}{2} \cos \left(\Omega_i + \Omega_j \right) t,$$
$$\sin \Omega_i t \cos \Omega_j t = \frac{1}{2} \sin \left(\Omega_i - \Omega_j \right) t + \frac{1}{2} \sin \left(\Omega_i + \Omega_j \right) t, \tag{81}$$

where δ_{ij} is the Kronecker symbol. The modal inertia force, $\{F_i\} = \{F_i\}^{(1)} + \{F_i\}^{(2)}$, is included in the left hand side of the modal differential equation (64), with an exchanged sign, and the extended equation is written in the matrix notation

31

$$\left[\tilde{N}_{ii}\right]\{\ddot{T}_j\}+\left[\tilde{M}_{ii}\right]\{\ddot{T}_j\}+\left[V_{ii}\right]\{\dot{T}_j\}+\left[\tilde{K}_{ii}\right]\{T_j\}$$
$$+\left[n_{ij}^{(4)}(t)\right]\{\ddot{T}_j\}+\left[n_{ij}^{(3)}(t)\right]\{\ddot{T}_j\}+\left[\tilde{m}_{ij}(t)\right]\{\ddot{T}_j\} \qquad (82)$$
$$+\left[\tilde{v}_{ij}(t)\right]\{\dot{T}_j\}+\left[\tilde{k}_{ij}(t)\right]\{T_j\}=\{F_0\sin\Omega_i t\},$$

where

$$\left[\tilde{N}_{ii}\right]=\left[N_{ii}\right]+\left[n_{ii}^{(04)}\right],\ \left[\tilde{M}_{ii}\right]=\left[M_{ii}\right]+\left[m_{ii}^0\right]-\left[n_{ii}^{(02)}\right],$$
$$\left[\tilde{K}_{ii}\right]=\left[K_{ii}\right]-\left[k_{ii}^0\right]+\left[n_{ii}^{(00)}\right],\ \left[\tilde{m}_{ij}(t)\right]=\left[m_{ij}(t)\right]-\left[n_{ij}^{(2)}(t)\right], \qquad (83)$$
$$\left[\tilde{v}_{ij}(t)\right]=\left[v_{ij}(t)\right]-\left[n_{ij}^{(1)}(t)\right],\ \left[\tilde{k}_{ij}(t)\right]=-\left[k_{ij}(t)\right]+\left[n_{ij}^{(0)}(t)\right],$$

and further

$$n_{ii}^{(04)}=\frac{PJ}{2gDS},\ n_{ii}^{(02)}=\frac{3PJ}{gDS}\Omega_i^2,\ n_{ii}^{(00)}=\frac{PJ}{2gDS}\Omega_i^4,\ n_{ij}^{(4)}(t)=\frac{PJ}{2gDS}f_{ij}^{(1)},$$
$$n_{ij}^{(3)}(t)=\frac{2PJ}{gDS}\Omega_j f_{ij}^{(2)},\ n_{ij}^{(2)}(t)=\frac{3PJ}{gDS}\Omega_j^2 f_{ij}^{(1)},\ n_{ij}^{(1)}(t)=\frac{2PJ}{gDS}\Omega_j^3 f_{ij}^{(2)},$$
$$\qquad (84)$$
$$n_{ij}^{(0)}(t)=\frac{PJ}{2gDS}\Omega_j^4 f_{ij}^{(1)},\ m_{ii}^0=\frac{P}{2gD},\ k_{ii}^0=\frac{P}{2gD}\Omega_i^2,$$
$$m_{ij}(t)=\frac{P}{2gD}f_{ij}^{(1)},\ v_{ij}(t)=\frac{P}{gD}\Omega_j f_{ij}^{(2)},\ k_{ij}(t)=\frac{P}{2gD}\Omega_j^2 f_{ij}^{(1)},$$

where

$$f_{ij}^{(1)}=\left(1-\delta_{ij}\right)\cos\left(\Omega_i-\Omega_j\right)t-\cos\left(\Omega_i+\Omega_j\right)t,$$
$$f_{ij}^{(2)}=\sin\left(\Omega_i-\Omega_j\right)t+\sin\left(\Omega_i+\Omega_j\right)t. \qquad (85)$$

As can be seen in (82) and (83), additional constant matrices (denoted by small letters) are added to the ordinary matrices (denoted by capital letters), while additional variable matrices, which represent parametric excitation, are separated. Both matrices are of the second order of magnitude characterized by the mass aspect ratio $m_{ii}^0/M_{ii} \approx P/(gml) = \varepsilon$. Due to that fact that the matrix modal equation (82) is split into a system of coupled modal equations

$$\tilde{N}_{ii}\dddot{T}_i + \tilde{M}_{ii}\ddot{T}_i + V_{ii}\dot{T}_i + \tilde{K}_{ii}T_i = F_0\sin\Omega_i t - \tilde{F}_i(t), \tag{86}$$

where

$$\tilde{F}_i(t) = \sum_{j=1}^{\infty}\left[n_{ij}^{(4)}(t)\ddddot{T}_j + n_{ij}^{(3)}(t)\dddot{T}_j + \tilde{m}_{ij}(t)\ddot{T}_j + \tilde{v}_{ij}(t)\dot{T}_j + \tilde{k}_{ij}(t)T_j\right], \tag{87}$$

is modal parametric force.

Now, the new formulation for beam vibrations due to moving inertia force can be compared with the known one presented in Section 2. It is obvious that the structure of modal equation (82) and (9) is the same. However, the new formulation deals with only two trigonometric functions (81), while the known one uses four functions, Eqs. (12). As a consequence, the new formulation and the known formulation include 8 and 12 time-dependent parametric matrices, respectively. Hence, the new formulation is quite a bit simpler than the known one, and this advantage of the modified Timoshenko beam theory is especially pointed out in the analytical solution of the problem, as can be seen further below.

8.3 Solution of linear vibrations

The solution of the modal differential equation (86) with ordinary excitation $F_0\sin\Omega_i t$ is of the same form as that shown in Section 5, Eq. (72), i.e.

$$\tilde{T}_i = \tilde{A}_i \left[\cos\Omega_i t - e^{-\gamma_i \tilde{\omega}_i t} \left(\cos\tilde{\omega}_i^* t + \gamma_i \frac{\tilde{\omega}_i}{\tilde{\omega}_i^*} \sin\tilde{\omega}_i^* t \right) \right]$$
$$+ \tilde{B}_i \left[\sin\Omega_i t - \frac{\Omega_i}{\tilde{\omega}_i^*} e^{-\gamma_i \tilde{\omega}_i t} \sin\tilde{\omega}_i^* t \right], \quad \tilde{\omega}_i^* = \tilde{\omega}_i \sqrt{1-\gamma_i^2}, \tag{88}$$

where according to (67) and (68)

$$\tilde{A}_i = \frac{\tilde{D}_A^i}{\tilde{D}_0^i}, \quad \tilde{B}_i = \frac{\tilde{D}_B^i}{\tilde{D}_0^i}, \tag{89}$$

and

$$\tilde{D}_A^i = -\Omega_i V_{ii} F_0,$$
$$\tilde{D}_B^i = \left(\Omega_i^4 \tilde{N}_{ii} - \Omega_i^2 \tilde{M}_{ii} + \tilde{K}_{ii} \right) F_0, \tag{90}$$
$$\tilde{D}_0^i = \left(\Omega_i^4 \tilde{N}_{ii} - \Omega_i^2 \tilde{M}_{ii} + \tilde{K}_{ii} \right)^2 + \Omega_i^2 V_{ii}^2.$$

The frequency $\tilde{\omega}_i$ is a natural frequency obtained from the homogenous part of Eq. (86), and according to (59) reads

$$\tilde{\omega}_i^{(1,2)} = \frac{1}{\sqrt{2\tilde{N}_{ii}}} \sqrt{\tilde{M}_{ii} \pm \sqrt{\tilde{M}_{ii}^2 - 4\tilde{N}_{ii}\tilde{K}_{ii}}}. \tag{91}$$

Considering the structure of the modal mass $\tilde{M}_{ii}$ and stiffness $\tilde{K}_{ii}$, Eqs. (82), the former is increased and the latter is decreased. As a consequence, the values of the natural frequencies will be reduced, and the amplitude of the forced vibration increased. Approximately $\tilde{\omega}_i = \omega_i (1-\varepsilon)$ and $\tilde{A}_i = A_i (1+\varepsilon)$.

8.4 Solution of subresonant parametric vibrations

The analytical solution of parametric vibrations is a rather complex task and is achieved by the perturbation method used in nonlinear dynamics [25]. Only the dominant terms of the time function $\tilde{T}_i$, Eqs. (88), which are in phase with the ordinary excitation $F_0 \sin \Omega_i t$, are taken into account for reasons of simplicity. Hence, one can write

$$
\begin{aligned}
\tilde{T}_j &= \tilde{B}_j \left(\sin \Omega_j t - \frac{\Omega_j}{\tilde{\omega}_j} \sin \tilde{\omega}_j t \right), \\
\dot{\tilde{T}}_j &= \tilde{B}_j \Omega_j \left(\cos \Omega_j t - \cos \tilde{\omega}_j t \right), \\
\ddot{\tilde{T}}_j &= \tilde{B}_j \Omega_j \left(-\Omega_j \sin \Omega_j t + \tilde{\omega}_j \sin \tilde{\omega}_j t \right), \\
\dddot{\tilde{T}}_j &= \tilde{B}_j \Omega_j \left(-\Omega_j^2 \cos \Omega_j t + \tilde{\omega}_j^2 \cos \tilde{\omega}_j t \right), \\
\ddddot{\tilde{T}}_j &= \tilde{B}_j \Omega_j \left(\Omega_j^3 \sin \Omega_j t - \tilde{\omega}_j^3 \sin \tilde{\omega}_j t \right).
\end{aligned}
\tag{92}
$$

By substituting formulae (84) and (92) into (87), one arrives at the following expression for the modal parametric force

$$
\begin{aligned}
\tilde{F}_i(t) = \frac{P}{2gD} \sum_{j=1}^{\infty} \tilde{B}_j \Omega_j^2 \Big\{ &a_j \left[f_{ij}^{(1)} \sin \Omega_j t - f_{ij}^{(2)} \cos \Omega_j t \right] \\
&+ b_j f_{ij}^{(1)} \sin \tilde{\omega}_j t - c_j f_{ij}^{(2)} \cos \tilde{\omega}_j t \Big\},
\end{aligned}
\tag{93}
$$

where

$$
\begin{aligned}
a_j &= -2 + 8 \frac{J}{S} \Omega_j^2, \quad b_j = \frac{\tilde{\omega}_j}{\Omega_j} + \frac{\Omega_j}{\tilde{\omega}_j} - \frac{J}{S} \frac{1}{\Omega_j \tilde{\omega}_j} \left(\Omega_j^4 + 6 \Omega_j^2 \tilde{\omega}_j^2 + \tilde{\omega}_j^4 \right), \\
c_j &= 2 - 4 \frac{J}{S} \left(\Omega_j^2 + \tilde{\omega}_j^2 \right).
\end{aligned}
\tag{94}
$$

35

The above terms without the ratio J/S are the result of matrices $m_{ij}(t)$, $v_{ij}(t)$ and $k_{ij}(t)$, and are dominant in the lower frequency domain, while the terms with J/S are obtained from the matrices $n_{ij}^{(k)}(t)$, $k = 1,2,3,4$, as an influence of rotary inertia.

Taking into account expressions (85), the products of trigonometric functions in (93) can be expanded by using the trigonometric formula

$$\sin\alpha\cos\beta = \frac{1}{2}\left[\sin(\alpha - \beta) + \sin(\alpha + \beta)\right].\tag{95}$$

As a result,

$$
\begin{aligned}
f_{ij}^{(1)}\sin\Omega_j t &= \frac{1}{2}\left[-\left(1-\delta_{ij}\right)\sin\alpha_{ij}^{(1)}t + \left(1-\delta_{ij}\right)\sin\alpha_{ij}^{(2)}t + \sin\alpha_{ij}^{(3)}t - \sin\alpha_{ij}^{(4)}t\right],\\
f_{ij}^{(2)}\cos\Omega_j t &= \frac{1}{2}\left[\sin\alpha_{ij}^{(1)}t + \sin\alpha_{ij}^{(2)}t + \sin\alpha_{ij}^{(3)}t + \sin\alpha_{ij}^{(4)}t\right],\\
f_{ij}^{(1)}\sin\tilde\omega_j t &= \frac{1}{2}\left[-\left(1-\delta_{ij}\right)\sin\alpha_{ij}^{(5)}t + \left(1-\delta_{ij}\right)\sin\alpha_{ij}^{(6)}t + \sin\alpha_{ij}^{(7)}t - \sin\alpha_{ij}^{(8)}t\right],\\
f_{ij}^{(2)}\cos\tilde\omega_j t &= \frac{1}{2}\left[\sin\alpha_{ij}^{(5)}t + \sin\alpha_{ij}^{(6)}t + \sin\alpha_{ij}^{(7)}t + \sin\alpha_{ij}^{(8)}t\right],
\end{aligned}\tag{96}
$$

where

$$
\begin{aligned}
\alpha_{ij}^{(1)} &= \Omega_i - 2\Omega_j,\ \alpha_{ij}^{(2)} = \Omega_i,\ \alpha_{ij}^{(3)} = \Omega_i,\ \alpha_{ij}^{(4)} = \Omega_i + 2\Omega_j,\\
\alpha_{ij}^{(5)} &= \Omega_i - \Omega_j - \tilde\omega_j,\ \alpha_{ij}^{(6)} = \Omega_i - \Omega_j + \tilde\omega_j,\ \alpha_{ij}^{(7)} = \Omega_i + \Omega_j - \tilde\omega_j,\\
\alpha_{ij}^{(8)} &= \Omega_i + \Omega_j + \tilde\omega_j.
\end{aligned}\tag{97}
$$

By substituting (96) into (93), one arrives at

$$\tilde F_i(t) = \frac{P}{4gD}\sum_{j=1}^{\infty}\sum_{k=1}^{8}\tilde B_j\Omega_j^2 d_{ij}^{(k)}\sin\alpha_{ij}^{(k)}t,\tag{98}$$

36

where

$$d_{ij}^{(1)} = -\left(2 - \delta_{ij}\right)a_j, \; d_{ij}^{(2)} = -\delta_{ij}a_j, \; d_{ij}^{(3)} = 0, \; d_{ij}^{(4)} = -2a_j,$$

$$d_{ij}^{(5)} = -\left(1 - \delta_{ij}\right)b_j - c_j, \; d_{ij}^{(6)} = \left(1 - \delta_{ij}\right)b_j - c_j, \; d_{ij}^{(7)} = b_j - c_j, \tag{99}$$

$$d_{ij}^{(8)} = -b_j - c_j.$$

The particular solution of Eq. (86) in the case of parametric excitation (98) is assumed in the harmonic form

$$T_i^{*P} = \sum_{j=1}^{\infty}\sum_{k=1}^{8}\left[A_{ij}^{(k)} \cos\alpha_{ij}^{(k)}t + B_{ij}^{(k)} \sin\alpha_{ij}^{(k)}t \right]. \tag{100}$$

By substituting (98) and (100) into (86) and equalizing the coefficients of the same harmonics, one obtains

$$A_{ij}^{(k)} = -\frac{D_{ij}^{A(k)}}{D_{ij}^{0(k)}}, \; B_{ij}^{(k)} = -\frac{D_{ij}^{B(k)}}{D_{ij}^{0(k)}}, \tag{101}$$

where

$$D_{ij}^{A(k)} = -\alpha_{ij}^{(k)}V_{ii}F_{ij}^{(k)},$$

$$D_{ij}^{B(k)} = \left[\left(\alpha_{ij}^{(k)}\right)^4 \tilde{N}_{ii} - \left(\alpha_{ij}^{(k)}\right)^2 \tilde{M}_{ii} + \tilde{K}_{ii}\right]F_{ij}^{(k)},$$

$$D_{ij}^{0(k)} = \left[\left(\alpha_{ij}^{(k)}\right)^4 \tilde{N}_{ii} - \left(\alpha_{ij}^{(k)}\right)^2 \tilde{M}_{ii} + \tilde{K}_{ii}\right]^2 + \left(\alpha_{ij}^{(k)}\right)^2 V_{ii}^2, \tag{102}$$

$$F_{ij}^{(k)} = \frac{P}{4gD}\tilde{B}_j\Omega_j^2 d_{ij}^{(k)}.$$

37

In order to enable satisfaction of the the initial conditions, the homogenous solution T_i^{*h} has to be added to the particular one, T_i^{*p}, forming in such a way total time function $T_i^* = T_i^{*p} + T_i^{*h}$. By setting $T_i^*(0) = 0$ and $\dot{T}_i^*(0) = 0$, one obtains similarly to (88)

$$
T_i^* = \sum_{j=1}^{\infty} \sum_{k=1}^{8} \left\{ A_{ij}^{(k)} \left[\cos \alpha_{ij}^{(k)} t - e^{-\gamma_i \tilde{\omega}_i t} \left(\cos \tilde{\omega}_i^* t + \gamma_i \frac{\tilde{\omega}_i}{\tilde{\omega}_i^*} \sin \tilde{\omega}_i^* t \right) \right] \right.
$$
$$
\left. + B_{ij}^{(k)} \left[\sin \alpha_{ij}^{(k)} t - \frac{\alpha_{ij}^{(k)}}{\tilde{\omega}_i^*} e^{-\gamma_i \tilde{\omega}_i t} \sin \tilde{\omega}_i^* t \right] \right\}.
$$

(103)

It is interesting to point out that the second and the third excitation harmonic in (98) have the ordinary forcing frequency Ω_i.

8.5 Solution of resonant parametric vibrations

8.5.1 Diagonal parametric time functions

In the case of $\Omega_i = \tilde{\omega}_i$, linear resonant vibrations occur, Section 8.3, and if $j = i$, coefficient $\tilde{B}_j$ in the parametric excitation (98), according to (89) and (90), takes zero or infinite value, depending on whether damping is included or not. This affects the time function (103) in the same way and none of the solutions is correct. Therefore, instead of applying the linear vibrations harmonic time function, Eqs. (92), one has to take the resonant form (73) into account, i.e.

$$\tilde{T}_i = A_{ri}\left(-\tilde{\omega}_i t \cos \tilde{\omega}_i t + \sin \tilde{\omega}_i t\right),$$

$$\dot{\tilde{T}}_i = A_{ri}\tilde{\omega}_i\left(\tilde{\omega}_i t \sin \tilde{\omega}_i t\right),$$

$$\ddot{\tilde{T}}_i = A_{ri}\tilde{\omega}_i^2\left(\tilde{\omega}_i t \cos \tilde{\omega}_i t + \sin \tilde{\omega}_i t\right), \tag{104}$$

$$\dddot{\tilde{T}}_i = A_{ri}\tilde{\omega}_i^3\left(-\tilde{\omega}_i t \sin \tilde{\omega}_i t + 2\cos \tilde{\omega}_i t\right),$$

$$\ddddot{\tilde{T}}_i = A_{ri}\tilde{\omega}_i^4\left(-\tilde{\omega}_i t \cos \tilde{\omega}_i t - 3\sin \tilde{\omega}_i t\right).$$

In the considered case, the trigonometric functions (85) are reduced to

$$f_{ii}^{(1)} = -\cos 2\tilde{\omega}_i t, \ f_{ii}^{(2)} = \sin 2\tilde{\omega}_i t, \tag{105}$$

and by substituting (104) and (105) into (87) and employing the formulae for the products of trigonometric functions like (95), one obtains the resonant parametric excitation

$$\tilde{F}_i(t) = F_i^0 \tilde{\omega}_i t \cos 3\tilde{\omega}_i t + G_i^0 \sin 3\omega_i t, \tag{106}$$

where

$$F_i^0 = -\frac{P}{gD} A_{ri}\tilde{\omega}_i^2\left(1 - 4\frac{J}{S}\tilde{\omega}_i^2\right), \ G_i^0 = 4\frac{PJ}{gDS} A_{ri}\tilde{\omega}_i^4. \tag{107}$$

The differential equation (86) with ignored damping and parametric excitation (107) is considered, and its particular solution is assumed in the same form as the excitation

$$T_{ii}^p = A_i \tilde{\omega}_i t \cos 3\tilde{\omega}_i t + B_i \sin 3\tilde{\omega}_i t. \tag{108}$$

By substituting (108) into (86) and equalizing the coefficients of the same functions on the left and right hand side, one obtains

$$A_i = \frac{F_i^0}{H_i}, \quad B_i = \frac{G_i^0 - 6\tilde{\omega}_i^2 \left(18\tilde{\omega}_i^2 N_{ii} - M_{ii}\right)}{H_i},$$

$$H_i = \left(3\tilde{\omega}_i\right)^4 N_{ii} - \left(3\tilde{\omega}_i\right)^2 M_{ii} + K_{ii} = 8\tilde{\omega}_i^2 \left(10\tilde{\omega}_i^2 N_{ii} - M_{ii}\right). \tag{109}$$

In order to satisfy the initial conditions, the homogenous solution of (86) is added to the particular one, i.e. $T_{ii} = T_{ii}^p + T_{ii}^h$, where $T_{ii}^h = C_i \cos \tilde{\omega}_i t + D_i \sin \tilde{\omega}_i t$. Hence, one finds from $T_{ii}(0) = 0$ and $\dot{T}_{ii}(0) = 0$ that $C_i = 0$ and $D_i = -A_i - 3B_i$. Finally, the diagonal parametric time function in resonance reads

$$T_{ii} = A_i \tilde{\omega}_i t \cos 3\tilde{\omega}_i t + B_i \sin 3\tilde{\omega}_i t - \left(A_i + 3B_i\right) \sin \tilde{\omega}_i t. \tag{110}$$

In practice, only the first resonance is of interest, i.e. $\Omega_1 = \tilde{\omega}_1$, and the total parametric time function of the first mode can be written as $T_1^* = T_{11} + \sum_{j=2}^{\infty} \sum_{k=1}^{8} T_{1j}^{(k)}$, where T_{11} is given by (110) for $i = 1$, and $T_{1j}^{(k)}$ is defined by the remaining terms of (103).

8.5.2 Off-diagonal parametric time functions

If $\Omega_j = \tilde{\omega}_j$, resonant vibrations are excited which affect all modes of index $i \neq j$ due to coupling. The time functions of the resonant ordinary vibrations due to moving gravity force, $\tilde{T}_j$, are presented by (104) with exchanged index i to j. Since $i \neq j$, $\delta_{ij} = 0$ and one can write for the trigonometric functions (85)

$$f_{ij}^{(1)} = \cos\left(\Omega_i - \tilde{\omega}_j\right)t - \cos\left(\Omega_i + \tilde{\omega}_j\right)t,$$
$$f_{ij}^{(2)} = \sin\left(\Omega_i - \tilde{\omega}_j\right)t + \sin\left(\Omega_i + \tilde{\omega}_j\right)t. \tag{111}$$

By substituting (111) into (84) and further into (87) together with (104) for $\tilde{T}_j$, one arrives at the modal parametric excitation forces

$$\tilde{F}_{ij}(t) = F_j^0 \tilde{\omega}_j t \cos\left(\Omega_i + 2\tilde{\omega}_j\right)t + G_j^0 \sin\left(\Omega_i + 2\tilde{\omega}_j\right)t$$
$$- F_j^0 \tilde{\omega}_j t \cos\left(\Omega_i - 2\tilde{\omega}_j\right)t + G_j^0 \sin\left(\Omega_i - 2\tilde{\omega}_j\right)t, \tag{112}$$

where F_j^0 and G_j^0 are specified by (107).

Accordingly, the differential equation (86) with ignored damping reads

$$\tilde{N}_{ii}\ddot{T}_{ij} + \tilde{M}_{ii}\ddot{T}_{ij} + \tilde{K}_{ii}T_{ij} = -F_j^0 \tilde{\omega}_j t \cos\alpha_{ij}t - G_j^0 \sin\alpha_{ij}t$$
$$+ F_j^0 \tilde{\omega}_j t \cos\beta_{ij}t - G_j^0 \sin\beta_{ij}t, \tag{113}$$

where $\alpha_{ij} = \Omega_j + 2\tilde{\omega}_j$ and $\beta_{ij} = \Omega_j - 2\tilde{\omega}_j$. The particular solution of (113) is assumed in the form

$$T_{ij}^p = A_{ij}\left(\tilde{\omega}_j t\right)^2 \sin\alpha_{ij}t + B_{ij}\tilde{\omega}_j t \cos\alpha_{ij}t + C_{ij}\sin\alpha_{ij}t$$
$$- X_{ij}\left(\tilde{\omega}_j t\right)^2 \sin\beta_{ij}t - Y_{ij}\tilde{\omega}_j t \cos\beta_{ij}t - Z_{ij}\sin\beta_{ij}t. \tag{114}$$

The expressions for the integration constants in (114) are derived in the Appendix. The complete solution of (113) is obtained by adding the homogenous solution, $T_{ij}^h = C_{ij}\cos\tilde{\omega}_j t + D_{ij}\sin\tilde{\omega}_j t$, to the particular one, i.e. $T_{ij} = T_{ij}^p + T_{ij}^h$. By satisfying the initial conditions $T_{ij}(0) = 0$ and $\dot{T}_{ij}(0) = 0$, one finds $C_{ij} = 0$ and

$$D_{ij} = -\frac{1}{\tilde{\omega}_i}\left[\tilde{\omega}_j\left(B_{ij} - Y_{ij}\right) + \alpha_{ij}C_{ij} - \beta_{ij}Z_{ij}\right].$$ (115)

Since only the first resonance is of practical interest, $\Omega_1 = \tilde{\omega}_1$, the total modal parametric time function can be written as $T_i^* = T_{i1} + \sum_{j=2}^{\infty}\sum_{k=1}^{8}T_{ij}^{(k)}$, $i = 2,3,...$, where T_{i1} is above as $T_{i1} = T_{i1}^p + T_{i1}^h$, and $T_{ij}^{(k)}$ is defined by the remaining terms of (103).

9 Beam vibrations after the passing of moving mass

The total time function of bending deflection due to moving gravity and inertia force is $T_{bi} = \tilde{T}_i + T_i^*$, Eqs. (88) and (103), and total beam deflection according to (62) reads

$$w = \sum_{i=1}^{\infty} \sin \frac{i\pi x}{l} \left\{ \left[1 + \left(\frac{i\pi}{l} \right)^2 \frac{D}{S} \right] T_{bi} + \frac{J}{S} \ddot{T}_{bi} \right\}. \tag{116}$$

When the moving mass reaches the beam end at the time instant t_P, the beam continues to vibrate in a natural way with initial values of modal displacements and a velocity equal to those at the end of the forced harmonic and parametric vibrations. Since the natural modes are the same in the forced and free vibrations, only the time function of the total deflection in (116) is considered, i.e.

$$T_{wi}(t_P) = \left[1 + \left(\frac{i\pi}{l} \right)^2 \frac{D}{S} \right] T_{bi}(t_P) + \frac{J}{S} \ddot{T}_{bi}(t_P). \tag{117}$$

The time function of the natural mode in the shifted time $\bar{t} = t - t_P$, introduced for reasons of simplicity, reads

$$T_i^h = \left(C_i \cos \tilde{\omega}_i^* \bar{t} + D_i \sin \tilde{\omega}_i^* \bar{t} \right) e^{-\gamma_i \tilde{\omega}_i \bar{t}}. \tag{118}$$

By satisfying the continuity conditions $T_i^h(0) = T_{wi}(t_P)$ and $\dot{T}_i^h(0) = \dot{T}_{wi}(t_P)$ one finds

$$C_i = T_{wi}(t_P), \quad D_i = \frac{1}{\tilde{\omega}_i^*} \left[\dot{T}_{wi}(t_P) + \gamma_i \tilde{\omega}_i T_{wi}(t_P) \right]. \tag{119}$$

10 Numerical examples

A simply supported thin-walled square tube with length $l = 100$ m, and cross-section $B \times H \times h = 10 \times 10 \times 0.1$ m, Fig. 1, is considered. The basic particulars are the following: cross-section area $A = 2(B + H)h = 4$ m^2, cross-section moment of inertia $I = (2/3)H^3 h = 66.667$ m^4, steel mass density $\rho = 8.65$ t/m^3, gravity constant $g = 9.81$ m/s^2, Young's modulus $E = 2.1 \cdot 10^8$ kN/m^2, shear modulus $G = E/(2(1 + v))$ and Poisson's ratio $v = 0.3$. The derived particulars are: distributed mass $m = \rho A$ and distributed mass moment of inertia $J = \rho I$. Moving gravity force is $P = C\rho g A H$, where C is a chosen coefficient. Moving mass is $P/g = C\rho A H$, or $P/g = 0$ if it is ignored. The shear coefficient for the thin-walled square tube is taken according to [26] as $k_S = \dfrac{20(1 - v)}{48 + 39v} = 0.4355$.

All numerical results are presented in a dimensionless form. Natural frequencies are normalized by the first natural frequency of the simply supported Euler-Bernoulli beam $\omega_{EB}^1 = (\pi/l)^2 \sqrt{EI/(\rho A)}$, so that the frequency parameter reads $\lambda_i = \pi^2 \omega_i / \omega_{EB}^1 = \omega_i l^2 \sqrt{\rho A/(EI)}$.

The velocity parameter is defined as the ratio of the moving mass velocity and the velocity which corresponds to the first natural mode, i.e. $\alpha = v_P/v_1 = \Omega_1/\omega_1 = \pi v_P/(\omega_1 l)$. Dimensionless time is $\tau = x_P/l = v_P t/l$. Beam deflection is normalized by the static value due to the lumped gravity force P acting at the midspan point of the beam, i.e. $\delta(x,t) = w(x,t)/w_P$, where $w_P = Pl^3/(48EI)$.

The influence of the moving mass on natural vibrations is investigated. The natural frequencies are determined as the eigenvalue of Eq. (86) by omitting damping, and the frequency parameter for three different moving mass values and the three values of the velocity parameter α are shown in Table 3. It is necessary to point out that natural frequencies do not depend on the position of the moving mass. The values

44

of the natural frequencies are reduced due to the increased mass and reduced stiffness for the first few natural modes.

Table 3. Influence of moving mass on natural frequency parameter $\lambda_i = \omega_i l^2 \sqrt{\rho A / EI}$ for simply supported beam, $l/H = 10$

α	Mode no.	$P/g = 0$	$P/g = \rho A h$	$P/g = 2 \rho A h$	$P/g = 3 \rho A h$
0.5	1	9.35	8.87	8.45	8.09
	2	32.89	31.74	30.70	29.76
	3	63.36	61.81	60.37	59.02
	4	96.46	94.87	93.35	91.88
	5	130.32	128.96	127.61	126.28
	6	164.24	163.25	162.25	161.23
	7	197.96	197.44	196.88	196.30
	8	231.43	231.41	231.36	231.28
	9	264.64	265.14	265.62	266.09
	10	297.63	298.66	299.68	300.71
1.0	1	9.35	8.62	8.03	7.54
	2	32.89	31.54	30.36	29.32
	3	63.36	61.69	60.16	58.75
	4	96.46	94.86	93.31	91.83
	5	130.32	129.06	127.78	126.49
	6	164.24	163.49	162.65	161.75
	7	197.96	197.82	197.53	197.15
	8	231.43	231.93	232.26	232.47
	9	264.64	265.80	266.77	267.63
	10	297.63	299.45	301.08	302.60
1.5	1	9.35	8.24	7.44	6.84
	2	32.89	31.24	29.92	28.80
	3	63.36	61.52	59.89	58.42
	4	96.46	94.85	93.28	91.77
	5	130.32	129.24	128.03	126.77
	6	164.24	163.86	163.21	162.41
	7	197.96	198.39	198.41	198.21
	8	231.43	232.71	233.47	233.95
	9	264.64	266.78	268.32	269.54
	10	297.63	300.64	302.96	304.95

The influence of the moving gravity force and inertia force on the beam response is analyzed via bending time functions. Ten natural modes are taken into account and the first three time functions determined for three values of the velocity parameter $\alpha = 0.5$, 0.9 and 1 by the ordinary procedure for the subresonant domain, Section 8.4, are shown in Figs. 2, 3 and 4. The governing differential equation (90) is also numerically integrated by employing the implicit Runge-Kutta method. Excellent agreement between the analytical and numerical results is obtained for the moving gravity force in the complete velocity range. In the case of moving inertia force, agreement is very good in the subresonant domain, Figs. 2 and 3, while in the resonant domain, $0.95 < \alpha < 1.05$, there are some discrepancies. They are at the maximum in resonance, $\alpha = 1$, as is evident in Fig. 4. Resonant response is determined by taking constant $\tilde{B}_1 = 0$ in the parametric excitation $\tilde{F}_i(t)$, Eq. (98), in order to avoid singularity, as explained in Section 8.5.1. If a special procedure for determining the resonant response from Section 8.5 is employed, excellent agreement between the analytical and numerical results is achieved as can be seen in Fig. 5. It is necessary to point out that the maximum realistic vehicle speed of $v_p = 300$ km/h = 83.3 m/s corresponds to the velocity parameter $\alpha = 0.133$, while the speed of sound in air $v_{sa} = 305$ m/s corresponds to $\alpha = 0.487$. For illustration, the bending and shear wave velocity in the Timoshenko beam, which is used in some references as a mathematical problem, seems to be unrealistic for ordinary engineering structures since $v_b = \sqrt{E/\rho} = 4927$ m/s and $v_s = \sqrt{k_s G/\rho} = 2016$ m/s, i.e. $\alpha_b = 59$ and $\alpha_s = 24$, respectively.

46

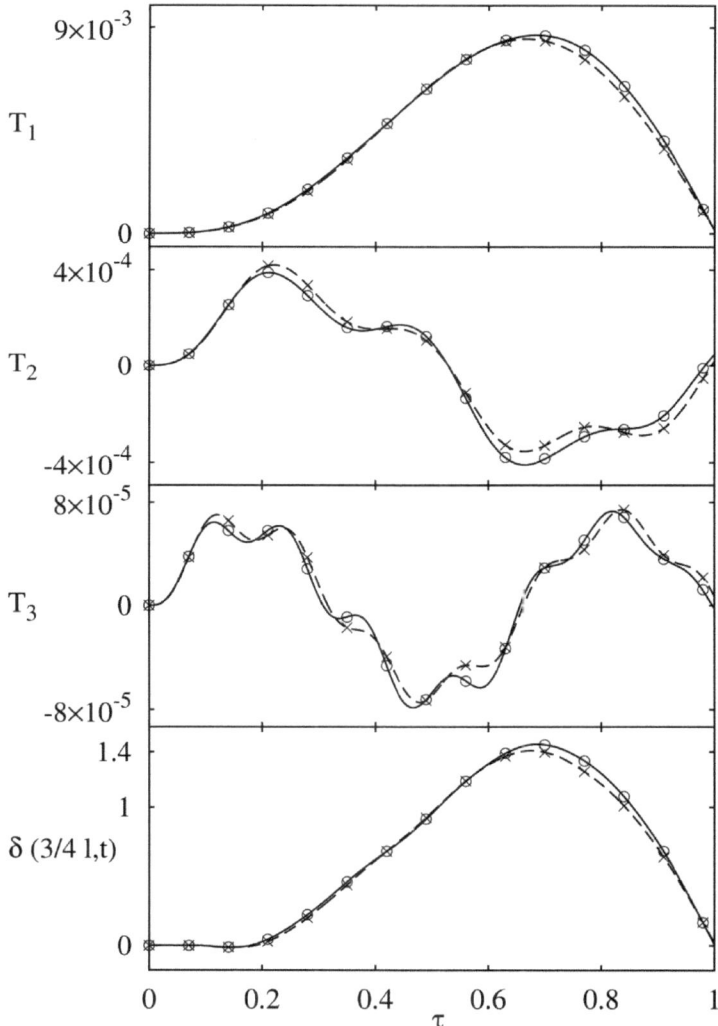

Fig. 2. Time functions of the first three natural modes and relative beam deflection, $l / H = 10$, $\alpha = 0.5$, $P/g = \rho A H$, $\gamma_i = 0.01$; $---$ gravity force, analytical; $\times \times \times$ gravity force, numerical; ———— gravity and inertia force, analytical; $\circ \circ \circ$ gravity and inertia force, numerical

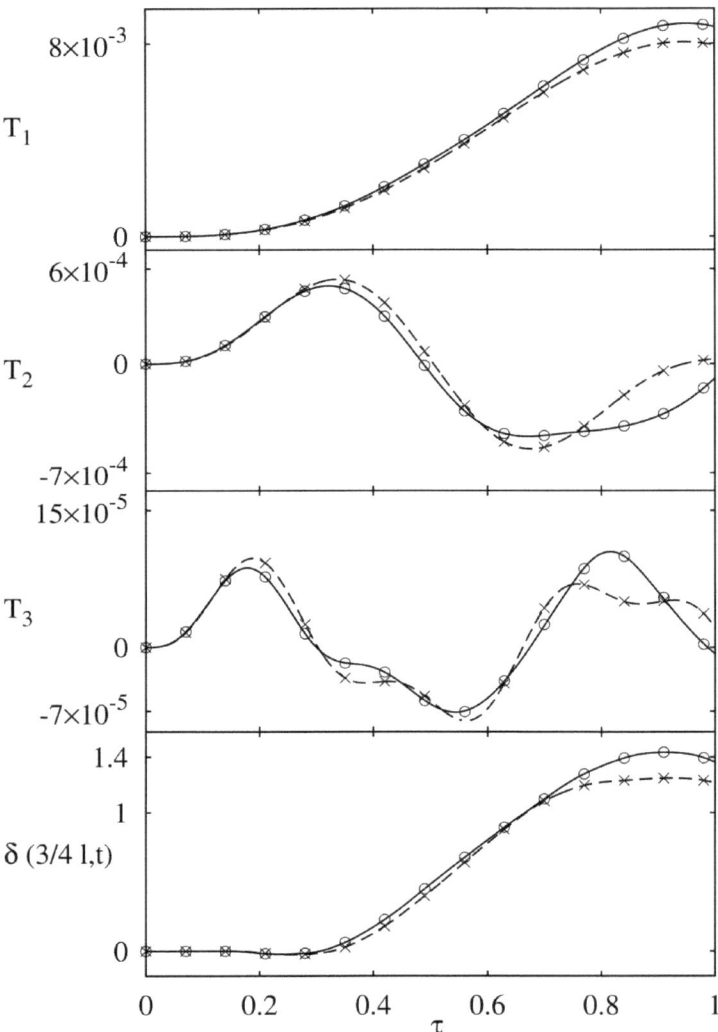

Fig. 3. Time functions of the first three natural modes and relative beam deflection, $l \, / \, H = 10$, $\alpha = 0.9$, $P/g = \rho A H$, $\gamma_i = 0.01$; $---$gravity force, analytical; $\times \times \times$ gravity force, numerical; —— gravity and inertia force, analytical; $\circ \circ \circ$ gravity and inertia force, numerical

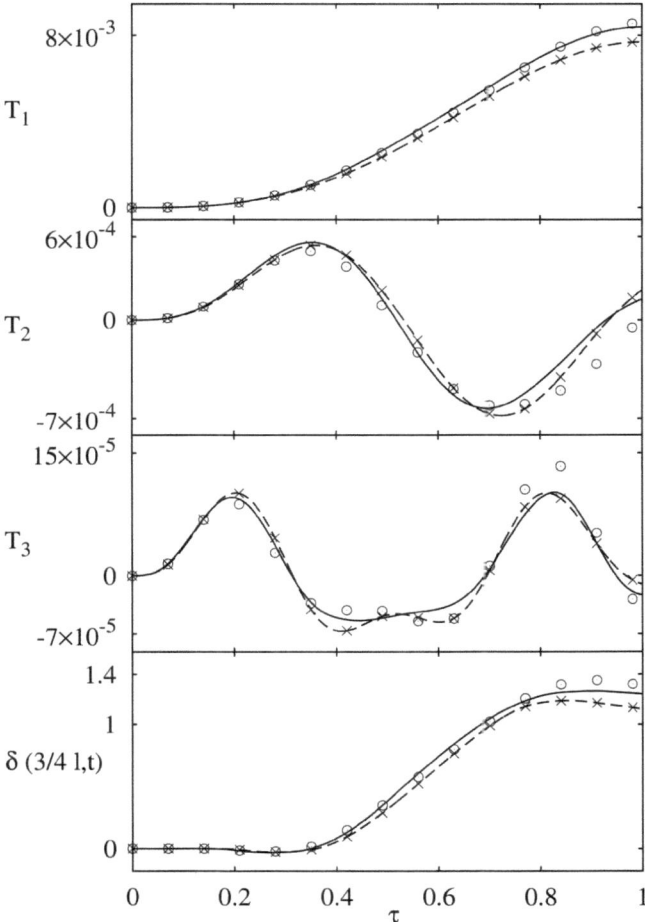

Fig. 4. Time functions of the first three natural modes and relative beam deflection in resonance, $l/H = 10$, $\alpha = 1.0$, $P/g = \rho A H$, $\gamma_i = 0.01$; $---$ gravity force, analytical; $\times \times \times$ gravity force, numerical; —— gravity and inertia force, analytical, approximate; $\circ \circ \circ$ gravity and inertia force, numerical

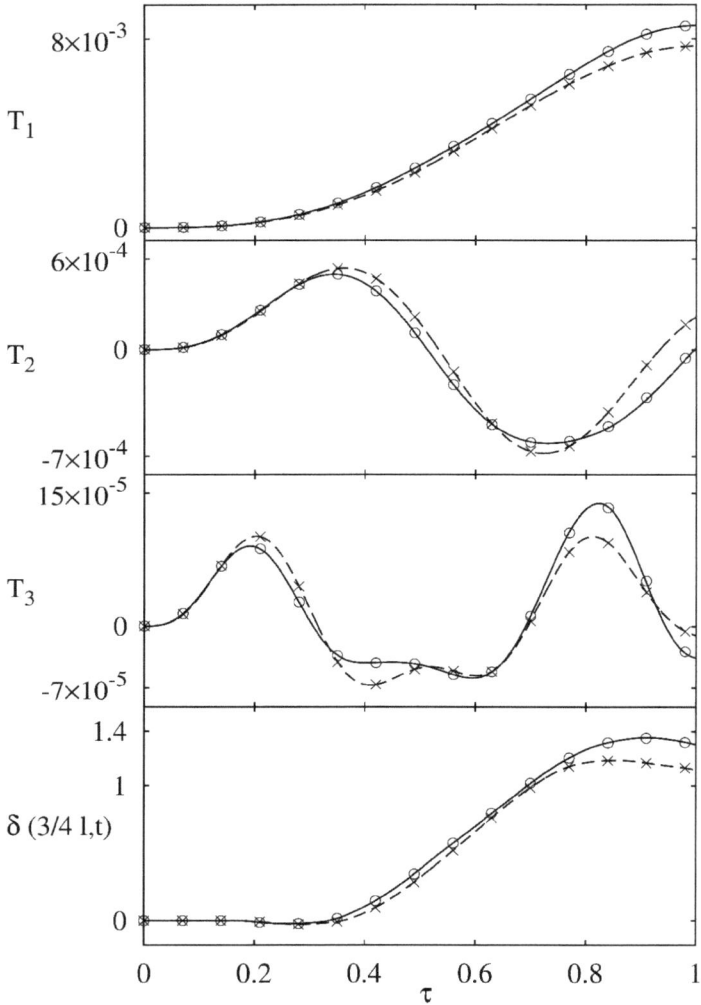

Fig. 5. Time functions of the first three natural modes and relative beam deflection in resonance, $l/H = 10$, $\alpha = 1.0$, $P/g = \rho A H$, $\gamma_i = 0.01$; $---$ gravity force, analytical; $\times \times \times$ gravity force, numerical; —— gravity and inertia force, analytical, exact; $\circ \circ \circ$ gravity and inertia force, numerical

In Fig. 6, the influence of the inertia force of different values of a moving mass on relative midspan deflection is presented in the time domain. The influence is increased in the second half of the passing time.

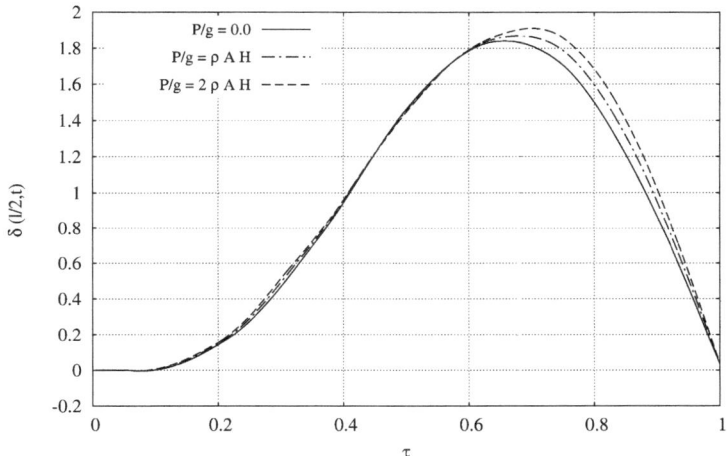

Fig. 6. Time history of relative midspan deflection, $l / h = 10$, $\alpha = 0.5$, $\gamma_i = 0.01$

Damping has quite a large and almost uniform influence on response in the whole velocity domain, as can be seen in Fig. 7, where the maximum midspan deflection is shown as a function of the velocity parameter.

Fig. 8 represents the moving mass trajectory for different values of the velocity parameter. If α is small, the trajectory is almost symmetric as in the case of a static load. The maximum value of the beam deflection below the moving mass is obtained at $\alpha \approx 0.4$ and then it is reduced approaching the resonance, $\alpha = 1$. Time histories of the relative beam deflection for the velocity parameter $\alpha = 0.1$ and 1 are shown in Figs. 9 and 10, respectively, where the moving mass trajectory is indicated. In the subresonant domain for lower values of the velocity parameter α, the beam is not deformed at the end of the moving mass passing time, $\tau \approx 1$, Fig. 9, since the inertia force is small. However, in the case of resonance, $\alpha = 1$, the beam deflection constantly grows with time, Fig. 10.

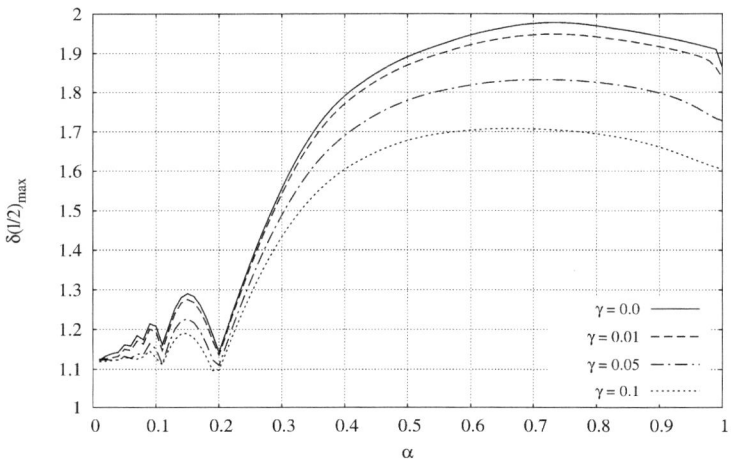

Fig. 7. Maximum value of relative midspan deflection, $l\,/\,H = 10$, $P/g = \rho A H$

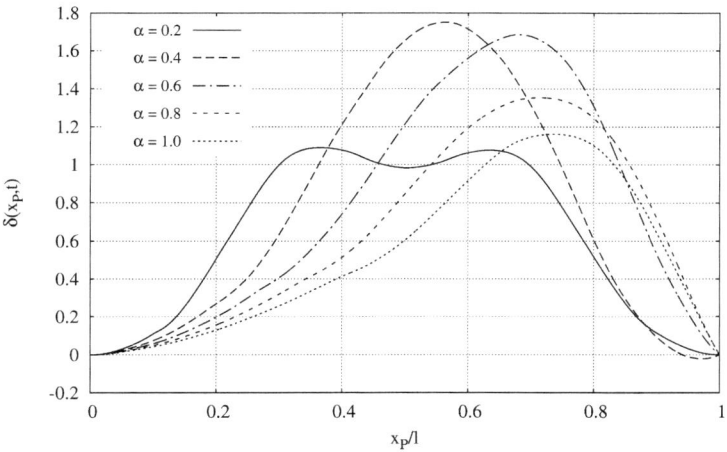

Fig. 8. Moving mass trajectory, $l\,/\,H = 10$, $P/g = \rho A H$, $\gamma_i = 0.01$

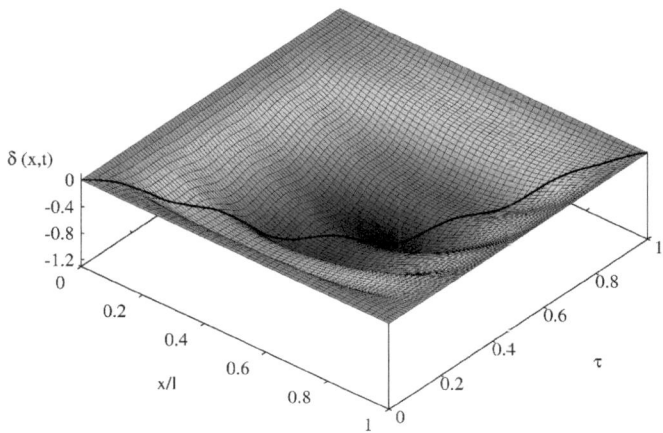

Fig. 9. Time history of relative deflection, $l/H = 10$, $P/g = \rho A H$, $\alpha = 0.1$, $\gamma_i = 0.01$

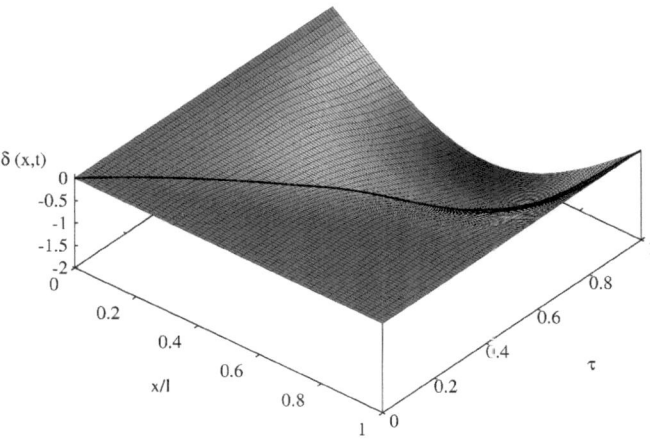

Fig. 10. Time history of relative deflection, $l/H = 10$, $P/g = \rho A H$, $\alpha = 1.0$, $\gamma_i = 0.01$

11 Conclusion

The analytical solution of the dynamic response of a beam-like structure exposed to a moving gravity and inertia force is a rather difficult task. The problem can be treated by the Timoshenko beam theory, which takes shear stiffness and rotary inertia into account. It deals with two coupled differential equations of motion with deflection and cross-section rotation as the main variables. That is not practical for an analytical solution, and therefore the Timoshenko beam theory is modified in such a way that deflection and rotation are decomposed into their constitutive parts and a single differential equation of the fourth order is derived in terms of bending deflection.

The problem of beam vibrations exposed to a moving force, formulated by a differential equation, cannot be solved directly due to moving discontinuity, and therefore the Galerkin method or energy approach is utilized. A systematic investigation performed in this book shows that the simplest formulation of the problem for an analytical solution is achieved by combining the modified Timoshenko beam theory with the Galerkin method or energy balance. The problem is solved by the modal superposition method.

Moving gravity force leads to a system of modal equations with constant coefficients, which is easily solved in the frequency domain. However, moving inertia force causes parametric excitation and the problem is solved by the perturbation method. The basic excitation harmonic causes a set of parametric harmonics, and the beam response is determined by the harmonic balance method. The obtained results are exact in the subresonant domain, while in the resonant domain some discrepancies with respect to the numerical solution are noticed. This problem is overcome by employing an exact resonant response function due to the gravity force for resonant parametric excitation. This rather complex procedure gives excellent results.

A beam exposed to the influence of moving force starts to vibrate from the rest and its response consists of steady state and transient functions, which are stable in resonance even if there is no damping. This is an explanation of why numerical time integration can be successfully performed in this specific case. Numerical examples show that damping has an almost uniform influence on the beam response in the whole moving mass velocity (frequency) domain.

An analytically performed parametric analysis not only manifests the influence of particular parameters on response, as a time integration solution, but also explains the reason for such behavior in a transparent way. An analytical solution for a beam with different boundary conditions is a rather difficult task, and therefore the numerical integration of differential equations of motion has to be undertaken. However, it is easier to evaluate the numerical results based on the knowledge achieved by an analytical solution. Analytically solved examples are at the same time educative and can be used as a benchmark for the numerical analysis of different types of beam-like structures exposed to a moving load.

Acknowledgment

This work was supported by the National Research Foundation of Korea (NRF) grant funded by the Korean Government (MEST) through GCRC – SOP (Grant No. 2011-0030669).

Appendix

Solution of a differential equation for off-diagonal parametric time functions

The governing differential equation (113) includes two excitation harmonics, and for reasons of simplicity only the first case is considered, i.e.

$$\tilde{N}_{ii}\ddot{\ddot{T}}_{ij} + \tilde{M}_{ii}\ddot{T}_{ij} + \tilde{K}_{ii}T_{ij} = -F_j^0 \tilde{\omega}_j t \cos \alpha_{ij} t - G_j^0 \sin \alpha_{ij} t. \tag{A1}$$

The particular solution of (A1) is assumed in the form

$$T_{ij}^p = A_{ij}\left(\tilde{\omega}_j t\right)^2 \sin \alpha_{ij} t + B_{ij}\tilde{\omega}_j t \cos \alpha_{ij} t + C_{ij} \sin \alpha_{ij} t. \tag{A2}$$

By substituting (A2) into (A1) and equalizing the coefficients of the same functions on the left and right hand side, one obtains the following algebraic equations

$$
\begin{aligned}
&\left(\alpha_{ij}^4 \tilde{N}_{ii} - \alpha_{ij}^2 \tilde{M}_{ii} + \tilde{K}_{ii}\right) A_{ij} = 0, \\
&4\tilde{\omega}_j \alpha_{ij}\left(-2\alpha_{ij}^2 \tilde{N}_{ii} + \tilde{M}_{ii}\right) A_{ij} + \left(-\alpha_{ij}^4 \tilde{N}_{ii} - \alpha_{ij}^2 \tilde{M}_{ii} + \tilde{K}_{ii}\right) B_{ij} = -F_j^0, \\
&2\tilde{\omega}_j\left(-6\alpha_{ij}^2 \tilde{N}_{ii} + \tilde{M}_{ii}\right) A_{ij} + 2\tilde{\omega}_j \alpha_{ij}\left(\alpha_{ij}^2 \tilde{N}_{ii} - \tilde{M}_{ii}\right) B_{ij} \\
&\qquad\qquad\qquad + \left(\alpha_{ij}^4 \tilde{N}_{ii} - \alpha_{ij}^2 \tilde{M}_{ii} + \tilde{K}_{ii}\right) C_{ij} = -G_j^0.
\end{aligned}
\tag{A3}
$$

There are two types of solutions of Eqs. (A3). If $\alpha_{ij}^4 \tilde{N}_{ii} - \alpha_{ij}^2 \tilde{M}_{ii} + \tilde{K}_{ii} \neq 0$, then $A_{ij} = 0$ and one finds

$$
\begin{aligned}
B_i &= \frac{F_j^0}{-\alpha_{ij}^4 \tilde{N}_{ii} - \alpha_{ij}^2 \tilde{M}_{ii} + \tilde{K}_{ii}}, \\
C_{ij} &= \frac{-G_j^0 - 2\tilde{\omega}_j \alpha_{ij}\left(\alpha_{ij}^2 \tilde{N}_{ii} - \tilde{M}_{ii}\right)}{-\alpha_{ij}^4 \tilde{N}_{ii} - \alpha_{ij}^2 \tilde{M}_{ii} + \tilde{K}_{ii}}.
\end{aligned}
\tag{A4}
$$

In the case that $\alpha_{ij}^4 \tilde{N}_{ii} - \alpha_{ij}^2 \tilde{M}_{ii} + \tilde{K}_{ii} = 0$, A_{ij} exists while C_{ij} is eliminated, and the remaining constants are obtained from the matrix equation

$$\begin{bmatrix} 4\tilde{\omega}_j \alpha_{ij}\left(-2\alpha_{ij}\tilde{N}_{ii} + M_{ii}\right) & -\alpha_{ij}^4 \tilde{N}_{ii} - \alpha_{ij}^2 \tilde{M}_{ii} + \tilde{K}_{ii} \\ 2\tilde{\omega}_j^2 \left(-6\alpha_{ij}^2 \tilde{N}_{ii} + M_{ii}\right) & 2\tilde{\omega}_j \alpha_{ij}\left(\alpha_{ij}^2 \tilde{N}_{ii} - M_{ii}\right) \end{bmatrix} \begin{Bmatrix} A_{ij} \\ B_{ij} \end{Bmatrix} = -\begin{Bmatrix} F_j^0 \\ G_j^0 \end{Bmatrix}. \quad (A5)$$

The solution of Eq. (113) for the second parametric excitation can be obtained in an analogous way.

If the mass moment of inertia J is ignored, the problem is quite simplified. The differential equation (113) reads

$$\tilde{M}_{ii}\ddot{T}_{ij} + \tilde{K}_{ii}T_{ij} = -\frac{P}{gD}A_{rj}\,\tilde{\omega}_j^2\left(\tilde{\omega}_j t \cos\alpha_{ij}t - \tilde{\omega}_j t \cos\beta_{ij}t\right). \quad (A6)$$

The integration constants in solution (114) are the following: if $\tilde{\omega}_i^2 - \alpha_{ij}^2 \neq 0$

$$B_{ij} = \frac{P}{gD}\frac{A_{rj}\tilde{\omega}_j^2}{\left(\tilde{\omega}_i^2 - \alpha_{ij}^2\right)\tilde{M}_{ii}}, \quad C_{ij} = \frac{2P}{gD}\frac{A_{rj}\tilde{\omega}_j^3 \alpha_{ij}}{\left(\tilde{\omega}_i^2 - \alpha_{ij}^2\right)^2 \tilde{M}_{ii}}, \quad (A7)$$

and if $\tilde{\omega}_i^2 - \alpha_{ij}^2 = 0$

$$A_{ij} = \frac{P}{4gD}\frac{A_{rj}\tilde{\omega}_j}{\tilde{M}_{ii}\alpha_{ij}}, \quad B_{ij} = \frac{P}{4gD}\frac{A_{rj}}{\tilde{M}_{ii}}\left(\frac{\tilde{\omega}_j}{\alpha_{ij}}\right)^2. \quad (A8)$$

The formulae for constants X_{ij}, Y_{ij} and Z_{ij} are obtained by inserting β_{ij} in (A7) and (A8) instead of α_{ij}.

References

[1]   S.P. Timoshenko, On the correction for shear of the differential equation for transverse vibration of prismatic bars. Philosophical Magazine, 41(6), 1921, pp. 744-746.

[2]   S.P. Timoshenko, On the transverse vibrations of bars of uniform cross-section. Philosophical Magazine, 43, 1922, pp.125-131.

[3]   I. Senjanović, Y. Fan, A finite element formulation of ship cross-sectional stiffness parameters. Brodogradnja, 41(1), 1993, pp. 27-36.

[4]   I. Senjanović, S. Tomašević, N. Vladimir, An advanced theory of thin-walled structures with application to ship structures. Marine Structures, 22(3), 2009, pp. 387-437.

[5]   M. Simsek, Forced vibration of an embedded single-walled carbon nanotube traversed by a moving load using nonlocal Timoshenko beam theory. Steel and Composite Structures, Vol. 11, No. 1, 2011, pp. 59-76.

[6]   K. Kiani, Q. Wang, On the interaction of a single-walled carbon nanotube with moving nanoparticle using nonlocal Rayleigh, Timoshenko, and higher-order beam theories. European Journal of Mechanics A/Solids 31, 2012, pp. 179-202.

[7]   S.P. Timoshenko, On the forced vibrations of bridges. Philosophical Magazine Series 6, 43, 1922, pp. 1018.

[8]   C.E. Inglis, A Mathematical Treatise on Vibration in Railway Bridges, Cambridge University Press, Cambridge, 1934.

[9]   P.K. Chatterjee, T.K. Datta, C.S. Surana, Vibrations of continuous bridges under moving vehicles. Journal of Sound and Vibration, 169, 1994, pp. 619-632.

[10]   L. Frýba, Dynamics of Railway Bridges, Thomas Telford Science Ltd., Prague, 1996.

[11]   E.C. Ting, J. Genin, J.H. Ginsberg, A general algorithm for moving mass problems. Journal of Sound and Vibration, 33(1), 1974, pp. 49-58.

[12]   M. Olson, On the fundamental moving load problem. Journal of Sound and Vibration, 154(2), 1991, pp. 299-307.

[13] L. Frýba, Vibrations of Solids and Structures under Moving Loads, Thomas Telford House, 1999.

[14] H.P. Lee, Transverse vibration of a Timoshenko beam acted upon by an accelerating mass. Applied Acoustics, 47, 1996, pp. 319-330.

[15] P. Sniady, Dynamic response of Timoshenko beam to a moving force. Journal of Applied Mechanics, 75(2), 2008, pp. 0245031-0245034.

[16] K. Kiani, A. Nikkhoo, B. Mehri, Prediction capabilities of classical and shear deformable beam models excited by a moving mass. Journal of Sound and Vibration, 320(2), 2009, pp. 632-648.

[17] G. Michaltsos, D. Sophianopoulos, A. N. Kounadis, The effect of a moving mass and other parameters on the dynamic response of a simply supported beam, Journal of Sound and Vibration, 191(3), 1996, pp. 357-362.

[18] M. Ichikawa, Y. Miyakawa, A. Matsuda, Vibration analysis of the continuous beam subjected to a moving mass. Journal of Sound and Vibration, 230, 2000, pp. 493-506.

[19] C.I. Bayer, B. Dyniewicz, Numerical Analysis of Vibrations of Structures and Moving Inertial Load, Springer-Verlag Berlin Heidelberg, 2012.

[20] B. Dyniewicz, C.I. Bayer, New feature of the solution of a Timoshenko beam carrying the moving mass particle, Arch. Mech., 62(5), 2010, pp. 327-341.

[21] I. Senjanović, N. Vladimir, Physical insight into Timoshenko beam theory and its modification with extension. Structural Engineering and Mechanics, 48(4), 2013, pp. 519-545.

[22] S.P. Timoshenko, Vibration Problems in Engineering, 2nd Edition, D. van Nostrand Company, Inc. New York, NY, USA, 1937.

[23] W.T. Thomson, Theory of Vibration with Applications. George Allen & Unwin, London, 1981.

[24] N.F.J. van Rensburg, A.J. van der Merve, Natural frequencies and modes of a Timoshenko beam, Wave Motion, 44, 2006. pp. 58-69.

[25] A.H. Nayfeh, Introduction to Perturbation Technique, Wiley-Interscience, New York, 1981.

[26] G.R. Cowper, The shear coefficient in Timoshenko's beam theory. Journal of Applied Mechanics, June 1966, pp. 335-340.

Printed by Books on Demand GmbH, Norderstedt / Germany